# Innovation Performance, Learning, and Government Policy

## The Economics of Technological Change
Edwin Mansfield, *General Editor*


*Growth, Innovation and Reform in Eastern Europe*
Stanislaw Gomulka

*The Semiconductor Business:*
*The Economics of Rapid Growth and Decline*
Franco Malerba

*Innovation Performance, Learning, and Government Policy:*
*Selected Essays*
Morris Teubal


During the past 25 years, economists have begun to study systematically the factors underlying the production of new technology. Among many other things, they have learned that the management of technological innovation entails a great deal more than the establishment of an R and D laboratory that turns out a lot of good technical output. A central problem facing a firm that attempts to be innovative is to effect a proper coupling between R and D, on the one hand, and marketing and production, on the other. For government agencies, there is a similar problem of integrating social and purely technological considerations.

This is the third volume in The University of Wisconsin Press Series on the Economics of Technological Change. Franco Malerba's *The Semiconductor Business* analyzed the decline of the European semiconductor industry relative to that in the United States and Japan. Stanislaw Gomulka's *Growth, Innovation and Reform in Eastern Europe* dealt with fundamental questions concerning the effects of economic systems on the rate of innovation. All of these books have attempted to understand why some organizations and societies are more successful than others in promoting and accepting new technologies, a question that will continue to be a central concern of future books in this series.

August 1986                                Edwin Mansfield
                                           Director, Center for
                                           Economics and Technology
                                           University of Pennsylvania

# Innovation Performance, Learning, and Government Policy

*Selected Essays*

Morris Teubal

ROBERT MANNING
STROZIER LIBRARY

JUL 2N 1983

Tallahassee, Florida

**The University of Wisconsin Press**

*Soc*
*HC*
*79*
*T4*
*T485*
*1987*

Published 1987

The University of Wisconsin Press
114 North Murray Street
Madison, Wisconsin 53715

The University of Wisconsin Press, Ltd.
1 Gower Street
London WC1E 6HA, England

Copyright © 1987
The Board of Regents of the University of Wisconsin System
All rights reserved

First printing

Printed in the United States of America

For LC CIP information see the colophon

ISBN 0-299-10950-X

**ROBERT MANNING
STROZIER LIBRARY**

JUL 29 1988

Tallanassee, Florida

To Eva

# Contents

# Figures

# Tables

# Preface

The role of innovation in the economic growth of firms, economic sectors, and countries is already well accepted. Although a number of different strands of research have dealt with this phenomenon, it is widely accepted that there is as yet no formal theory which is sufficiently general and useful for explaining the causes and effects of innovation. Among the best-known approaches we may mention the growth accounting approach pioneered by Abramovitz, Schmookler, and Solow, and later developed by Griliches and others. Recent theoretical work has focused on another topic—the so-called Schumpeterian competition and related links between market structure and innovation. Important theoretical and empirical contributions in these areas have been made in the last two decades by E. Mansfield, A. Phillips, R. Nelson, S. Winter, J. Stiglitz, and others.

A somewhat different tradition underlies the papers collected in this book. The starting point is the nature of the innovation process and the associated learning processes in the production and diffusion of innovations. This tradition originated with the innovation studies conducted in the sixties and early seventies by C. Freeman and associates, E. Mansfield, R. Nelson, N. Rosenberg, and others. These and subsequent studies have enhanced our conceptualization of the innovation process and have clarified our notions of "technology" and "needs" beyond the usually accepted economic theory notions of production function and demand function. Useful concepts such as focusing devices, natural trajectories, need determination, and technological paradigms have emerged. An important contribution to this process came from microeconomic case studies of the technological evolution, the learning processes, and the innovation patterns of firms and economic sectors, both in advanced and in developing countries. These case studies have also helped to distinguish between production and investment engineering capabilities and between experience-based knowledge and knowledge resulting from R & D, and to determine the conditions required to efficiently accumulate knowledge and the different kinds of knowledge being accumulated by different economic sectors. They are beginning to influence policy and our way of thinking in such a manner that may eventually lead to a relevant theory of innovation. This book should be viewed as a modest attempt to make a contribution to this process. It collects a number of case studies of firms in the fields of electronics, metal-working, and capital goods, and several other papers, concerned with various theoretical, methodological, and policy issues. The papers, which were written during the last twelve years, are organized into three main sections: Success and Failure in Innovation (part 1); Learning, Technological Capabilities, and Spin-offs (part 2); and Government Policy, Externalities, and Economic Growth (part 3). Each part is preceded by an introduction relating the various papers to the main issues arising in the relevant literature.

# Acknowledgments

The first papers were written at the Maurice Falk Institute for Economic Research in Israel, during the years 1973–1977. I am particularly indebted to the institute and to its director at the time, Michael Bruno, for encouragement and help in those early days of "muddling through." Christopher Freeman and the late Simon Kuznets also provided useful suggestions and vital encouragement. I had the privilege, during those years, of being associated with Manuel Trajtenberg in the design and execution of some of the early work. My evolving approach to the subject and the continuous conceptual clarification achieved owe a great deal to our interaction and mutual understanding. Richard Nelson has, since 1977, been a constant source of advice, encouragement, and suggestions. My later attempts at relating the microeconomic issues to broader policy questions were very much influenced both by his work and by his comments. I particularly appreciate the encouragement given at critical moments in the evolution of my work. Edwin Mansfield and Keith Pavitt, on their part, showed continued interest in my research, a fact which also reinforced my determination to continue with my line of thought.

It is difficult to trace my debt to the numerous other individuals with whom I interacted in the course of writing the papers of this volume. Thanks are due to Larry Westphal, S. Freund, Jorge Katz, Roy Rothwell, Nathan Rosenberg, Zvi Griliches, Yoram Ben-Porat, Danny Tsiddon, Philip Maxwell, Edward Steinmueller, Pablo Spiller, Carl Dahlman, and others. During the last two years I benefited from collaborating with Moshe Justman on a number of topics including the attempt to relate innovation studies with economic theory. My thanks go to him and to Beni Toren and to the Jerusalem Institute for Israel Studies for the sponsorship and finance of our current collaborative research effort.

Chapters from this book have appeared earlier in different form. Chapter 1 appeared in *Research Policy* 5 (1976): 354–79. Chapter 2 appeared in *Re-*

*search Policy* 6 (1977): 254–75. A shorter version of chapter 3 was published in *OMEGA* 5, no. 4 (1977). Chapter 4 appeared in *Industrial Innovation: Technology, Policy, Diffusion*, M. Baker, ed. (London: Macmillan, 1979), pp. 266–93. Chapter 5 was published in *World Development* 12 (1984): 849–65. Chapter 6 was published in *Research Policy* 11 (1982): 333–46. Chapter 7 appeared in *The Problem with Technology*, D. Lamberton et al., eds. (London: Pinter, 1984). Chapter 8 appeared in *Minerva* 21 (1983): 172–79. A shorter version of chapter 9 is to appear in *Research Policy*. Chapter 10 was published in *Research Policy* 11 (1982): 271–87. Chapter 11 was published in *The Economics of Relative Prices*, B. Czikos-Nagy, D. Hague, and G. Hall, eds. (London: Macmillan, 1984): 117–39. Chapter 12 appeared in *Trade, Stability, Technology, and Equity in Latin America* (New York: Academic Press, 1982), chapter 14.

# 1 Success and Failure in Innovation

# Introduction to Part I

The increased importance of technological innovation for economic growth has stimulated in the last two decades a number of attempts to analyze the causes of success and failure in innovation (and the characteristics differentiating successful from other innovations). In order to evaluate the methodology used in the various studies, Mowery and Rosenberg (1979) have conducted an important critical survey of the field and the conclusions obtained. After reviewing a number of different studies, they point out: "A final pair of studies is noteworthy for their attempts to compare directly successes and failures in innovation. Both SAPPHO from the Science Policy Research Unit at the University of Sussex and FIP—Falk Innovation Project—from the Maurice Falk Institute for Economic Research in Israel, compare innovations which failed either before (i.e., research is terminated) or following introduction." The FIP studies are the first two chapters in this volume. I will first try to set these within the development of innovation performance studies in general. I will then critically assess the conclusions of the studies and some implications. Finally, I will end this introduction with a discussion of some outstanding remaining issues that have emerged from the success-failure literature.

### Survey of Innovation Performance Studies

A central role in the evaluation of the field was played by project SAPPHO (Science Policy Research Unit, University of Sussex 1972; Rothwell et al. 1974). This project represent the first systematic attempt to compare successful and failed innovations by means of a common methodology (the paired-comparison method) and a common set of variables of potential significance in innovation performance. In its final stage it considers a set of 43 innovation pairs, 22 significant or "breakthrough" instrument innovations and 21 significant chemical innovations. The instruments are new products used in industry (e.g., digital voltmeter) and/or research (e.g., scanning electron microscope), while most chemical innovations involve new processes, mostly of standard heavy chemicals like ammonia, acetylene, methanol, urea. Each pair is composed of a commercially successful innovation *and* a commercially failed one belonging to the same family or type, where both successes and failures are actually launched commercially. The study considers over 100 variables which previous literature considered of potential interest. Each of the variables considered was applied to the set of pairs in order to see whether it differed between the successful element of a pair and the failed element of the same pair, across a significantly large number of pairs. Thus, there is a comparison within and not across innovation pairs. The conclusions involve both general conclusions on the whole set of innovations and area-specific ones in relation to instruments and chemical innovations (interindustry comparisons). The main general conclusions are: successful innovators (relative to failed innovators) have a much better understanding of user needs; they pay more attention to marketing, perform their development work more efficiently but not necessarily more quickly, and make more effective use of outside technology and scientific advice; and the responsible individuals are usually more senior and have greater authority than their counterparts that fail. The most interesting results, in my opinion, concern the interindustry differences in the factors associated with success. They tend to show that in instruments the "constraints" of demand and marketing are of paramount importance relative to their importance for chemical innovations. In a significant number of cases successful instrument innovations relative to failed ones are launched later, have longer development lead times, and represent a less radical technology from the firm's point of view; and their developers understand user needs better and undertake a larger marketing effort.

Before proceeding to the FIP studies included in this volume we should mention some additional contributions to the field. A SAPPHO-type study conducted by Maidique and Zirgen (1984) discusses the determinants of new-product success within a sample of 158 products of the United States electronic industry, 50% of which were commercial successes. The results obtained confirm in general terms the results of SAPPHO and those obtained by other researchers (see Cooper 1983): the importance of in–depth understanding of the customer and the market place; of well-coordinated design, production, and marketing functions; of drawing from existing technological and marketing strengths of the developing business units; of proficiency in marketing and in research and development, etc. Some of these as well as other factors are included in Mansfield et al.'s econometric study of the role of various organizational and strategic factors in a firm's probability of success in an R & D program composed mostly of development projects (basic research, for obvious reasons, was not included) (see Mansfield et al. 1977, chap. 2). Their results are based on data obtained from 20 major firms in the chemical, drug, petroleum, and electronics industries. An interesting feature of the study is the distinction made between three probabilities: the probability of technical completion; the probability of commercialization (given technical completion); and the probability of economic success (given commercialization). These probabilities indicate the percentage of R & D dollars that went for completed or successful projects. Intensive interviewing enabled the researchers to obtain estimates for each one of these probabilities for most of the 20 firms. The product of all three probabilities equals the probability that an R & D project begun by the firm will be an economic success. Through testing of a simple econometric model relating each one of these probabilities to a number of factors, the authors show, among other results: (1) all three probabilities increase the less resources are spent in the R & D projects prior to a serious study or project evaluation from the point of view of potential market and profits; (2) the probabilities of technical completion and of commercialization are higher in firms that devote more of their R & D resources to "demand-pull" projects rather than to "technology-push" projects. These results show the importance of integrating technological considerations with economic considerations early in the game (a fact confirmed by other studies as well). Additional work of the authors shows how reorganization of a firm leading to a greater integration between R & D and marketing increases the probability of commercialization (given technical completion).

The FIP studies included in this book are inspired by SAPPHO although they differ significantly from it both in objectives and in methodology. The objective is to analyze the pattern of revealed comparative advantage within a particular segment of the Israeli electronics industry—the biomedical electronics instrumentation sector (23 innovations, of which 5 were successes and 18 were failures[1])—that is, in what "areas" were the commercial successes concentrated? In contrast, SAPPHO did not really touch upon comparative advantage, and this was a direct outcome of its paired-comparison methodology. Each success is compared with the failure belonging to the same pair, i.e., belonging to the same product class, and not with failures (or successes) belonging to other pairs. What therefore differentiates the failures from successes in the SAPPHO study are a number of characteristics of the innovation, i.e., tactics or aspects of the management of innovation such as marketing, rather than the product areas themselves.

Our review of the main conclusions, shows, first, that all the successes were

---

1. This sample represented practically the whole "universe" of innovations in the sector till the beginning of 1973 (14 other innovations which were still in process were not included in the analysis).

products directed to users who had already translated (or defined) their needs in terms of product class and product functions ("I need an electronic cardiac monitoring system with the following parameters: ECG, heart rate, pulse rate," etc.)—this was termed a situation where users had a relatively high degree of need or market determinateness (see chap. 4 of this volume). Second, the launching of products possessing characteristics matching these needs was only a necessary, but not a sufficient, condition: another requirement for success was the existence of a significant increase in the efficiency of those functions desired by the user, even at the expense of a higher price. Third, failures either involved completely new products (for which users had undefined needs) or did not involve sufficient increases in functional efficiency of the more standard products; they generally resulted from inadequate project selection rather than from inadequate project execution (the latter was found to be only a consequence of the former). Inadequate project selection resulted from an inadequate estimation of the difficulties of marketing new products (the "acceptance problem") and correspondingly a lack of awareness of the marketing (including user involvement) requirements for success. Thus "large" projects were regarded as "small" and therefore were undertaken in some cases by small firms, which encountered in the process a financial constraint leading to project suspension (in some cases even before product launch). Entrepreneurs regarded R & D and marketing as interchangeable: "The product is so good it sells by itself." This is true with respect to R & D which does not involve additional novelty to users, i.e., the R & D which increases the efficiency of existing functions. It is not true with respect to R & D which leads to completely new and unknown products. In this case, the more R & D, the more marketing is required. These conclusions are largely consistent with SAPPHO's for instruments.

### Assessment of Results and Some Implications

The results of SAPPHO-related studies demonstrate the importance of the role of demand factors in explaining the commercial performance of innovations. While Schmookler (1966) emphasized market size as the main determinant of the distribution of patents on capital goods (innovations) across the various industrial branches using these goods, SAPPHO-related studies emphasize the importance of a thorough understanding of users' needs on the part of the firm and of significant and proficient marketing in explaining the commercial performance of innovation. The FIP studies attempted to identify within which *classes* of goods product innovation was likely to succeed commercially in the particular circumstances of the Israeli biomedical electronics sector of the late sixties and early seventies. The answer lay in those classes which expressed or translated the needs of a clearly specified class of users (their "needs" were highly determinate). Under these circumstances both the user acceptance risk and the innovator's risk of misunderstanding user needs would be relatively low, and so the marketing effort required to overcome both these risks would be relatively less. These results receive strong indirect support from the work of von Hipple (1976, 1978) on the role of users in the innovation process. Von Hipple shows convincingly that users may play an active role beyond the mere transmission of need signals to innovator firms. Their role may involve building prototypes of the new products, testing and using them, and transmitting the information to other users and to potential suppliers. This customer-active paradigm has been found to be of extreme importance in a number of industrial sectors. While von Hipple's work originally referred to scientific instruments and certain machinery areas, it is clear since then that the importance of his customer-active paradigm refers to a wide variety of areas within the machinery and instruments sectors (see Pavitt 1984). In this connection it is interesting to note that FIP covered biomedical electronics instrumentation, while SAPPHO's results on the importance of user needs were particularly clear in relation to its sample of instrument innovations.

The above results of the innovation performance studies should not necessarily be interpreted as establishing the "dominance" of demand factors over supply factors in explaining innovation performance (see Mowery and Rosenberg 1979). They can however be used in order to understand the relative importance of demand (and marketing) factors in different industries. The above authors state, in relation to the concept of need determination: "In all of the extensive discussions of the role of demand in the innovation process which have been surveyed here, this concept stands out as a useful one for comparing the role and importance of 'demand-pull' forces in different industries, something which cannot be done merely using the broad, fuzzy concepts of 'user needs' or 'need recognition'." In this connection it is useful to recall some of the differences recorded in the SAPPHO study between the factors explaining success within instruments and those explaining success within the set of chemical innovations. Need, market, and marketing-related variables are likely to be extremely important for machinery, instruments, and systems relative to their importance in other sectors.

### Outstanding Research Issues

A number of issues and themes requiring further research and clarification are suggested from the innovation performance literature. I will briefly refer to some of them only.

### The need for a theory of needs

Since an innovation is best defined as a "coupling" between technological knowledge and needs, advances in conceptualizing the phenomenon of innovation depend on advances in our notions about either one or both of the underlying forces. While the need concept used in economics—market demand—is useful for helping explain the rate of output of a particular good, it is useful only for explaining the incentives to introduce a limited group of innovations—process innovations and some kinds of product innovations. For example, when the need can be expressed in terms of concrete product characteristics, then the demand for these characteristics represents the market incentive for the innovation. But this case also presupposes the prior existence of a related product. The concept of market demand is not, however, *generally* useful for explaining the need for an innovation. Our notion of market or need determination is hopefully one step in the direction of developing a theory of needs. These ideas are discussed in chapter 1 and further developed in chapter 4.

### Methodological issues in innovation performance research

Chapter 3 suggests the grouping of the various explanatory factors affecting innovation performance into one of three possible levels: level I, which includes what may be termed "manifestations of performance" such as the degree of understanding of user needs on the part of the innovator; level II, firm behavior variables such as the amount of expenditures in marketing; and level III, firm characteristics such as size. While level I variables directly affect performance, the other variables have an effect only through their impact on level I variables. This suggests the possibility of generating *subsystems* of interrelated variables to analyze innovation performance, rather than correlating performance with (apparently) independent variables. Evidence from a pilot study reported in chapter 3 suggests very strongly that a *user needs subsystem* of variables offers a better explanation than other subsystems of variables for the commercial performance of SAPPHO's subset of instrument innovation pairs. Clearly, paired-comparison-type innovation studies could benefit from introducing a hierarchy of the variables potentially affecting performance. There is ample room for further developing and applying these hierarchies.

### The learning perspective to innovation performance

All of the studies reviewed up to now were "static"; i.e., they were concerned with explaining the pattern of commercial performance of a set of nondated innovations. In these studies commercial performance was related to the *direct* profitability flowing from sales of the innovation concerned. In reality, innovations contribute commercially beyond their direct profitability, i.e., via the economic value of the spin-offs they contribute to subsequent innovations of the firm. Thus a learning or dynamic perspective is required in order to truly discuss innovation performance (e.g., a static "failure" may well set the stage for a string of future commercial successes). No such studies are yet available, since a number of preliminary issues have yet to be resolved; e.g., what are the relevant spin-offs and how can we measure their economic value? Part II deals explicitly with some of these preliminary issues, including in chapter 6 a specific study of the chain of interrelated R & D projects of a successful Israeli instruments firm. Once a number of such studies are performed, it may be possible to think about designing a "dynamic" innovation performance study.

### References

Cooper, R. C. 1983. "A Process Model for Industrial New Product Development." *IEEE Transactions on Engineering Management* EM-30 (1) (November): 2–11.

Freeman, C. 1965. "Research and Development in Electronics Capital Goods." *National Institute Economic Review* 34 (November): 40–91.

Maidique, M. A. and B. J. Zirgen, 1984. "A Study of Success and Failure in Product Innovations: The Case of the U.S. Electronics Industry." *IEEE Transactions on Engineering Management* EM-81 (4) (November): 192–203.

Mansfield, E., et al. 1977. *The Production and Application of New Industrial Technology*. New York: W. W. Norton & Company.

Mowery, D., and N. Rosenberg. 1979. "The Influence of Market Demand upon Innovation: A Critical Review of Some Recent Empirical Studies." *Research Policy* 8: 102–53. Reprinted in N. Rosenberg. *Inside the Black Box: Technology and Economics*. Cambridge: Cambridge University Press, 1982.

Pavitt, K. 1984. "Sectoral Patterns of Technical Change: Towards a Taxonomy and Theory." *Research Policy* 13: 343–73.

Rothwell, R., C. Freeman, A. Horsley, V. T. P. Jervis, A. B. Robertson, and J. Townsend. 1974. "SAPPHO Updated—Project SAPPHO Phase II. *Research Policy* 3 (November): 258–91.

Schmookler, J. 1966. *Invention and Economic Growth*. Cambridge: Harvard University Press.

Science Policy Research Unit, University of Sussex. 1972. *Success and Failure in Industrial Innovation: Report on Project SAPPHO*. London: Centre for the Study of Industrial Innovation.

Teubal, M. 1978. "Threshold R & D Levels in Sectors of Advanced Technology." *European Economic Review*, 7:395–402.

von Hippel, E. 1976. "The Dominant Role of Users in the Scientific Instrument Innovation Process," *Research Policy*. 5:212–39.

von Hippel, E. 1978. "A Customer-Active Paradigm for Industrial Product Idea Generation." *Research Policy* 7:240–66.

# Performance in Innovation in the Israeli Electronics Industry
## A Case Study of Biomedical Electronics Instrumentation

## Morris Teubal, Naftali Arnon, and Manuel Trachtenberg

The purpose of our project is to develop a framework for understanding performance in innovation of science-based industry. Such a framework would enable us to define optimum strategies for science-based firms and optimum patterns of government support to science-based industry. We here report some preliminary findings on a case study of the Israeli biomedical electronics sector.

The methodological approach of the report consists in uncovering the components of optimum firm strategy from the simultaneous observation of performance in innovation at the level of individual R & D programs and the actual firm strategies which led to that performance. The research is therefore a combination of both empirical and theoretical research: there is no formal-theoretical model to determine from the outset the nature of the empirical test—the empirical and theoretical aspects continuously influence each other, and both are outputs of the project.

For our purpose we can distinguish two levels of firm strategy: an overall level and a program level. The overall strategic decisions[1] of a firm refer to its choice of areas of activity and to the extent of involvement in each of them. At the R & D program level, on the other hand, the strategic decisions refer to the characteristics of the program (the innovation profile, the degree of offensiveness, etc.). The conclusions of this report refer to both types of strategic decision, although somewhat more to those related to R & D programs.

The research will be in three stages. Stage 1 is to identify the variables most significantly related to performance; stage 2 is an analysis of failure; and stage 3 deals with constraints and policy. Each stage adds a dimension of its own to the analysis of the link between performance and firm strategy. This report

---

1. On the concept of strategic decision see Ansoff 1968, chap. 1.

covers only part of stage 1, and its conclusions should therefore be regarded as preliminary. At stage 3 we shall attempt to link government policy to performance in order to arrive at some useful policy recommendations.

### Relation to Previous Work

This research resembles the work of Myers and Marquis (1969), the Centre for the Study of Industrial Innovation (1971), and the report on project SAPPHO (Science Policy Research Unit, University of Sussex, 1972) in that the unit of analysis is the individual innovation or R & D program. It also is related to the research done by Jones (1969) in that it is specific to biomedical electronics and attempts to cover as much of the industry as possible. Of these, our report most strongly resembles the report on project SAPPHO not only because of the similarity in the unit of analysis but also because it includes both successes and failures. However, while owing a great deal to SAPPHO and to Freeman's previous work (1965), it is different in several respects, some of which we now mention briefly.

1. The SAPPHO report studies 29 innovations in the chemical and instruments sector; for each innovation there is a pair composed of one successful and one failed R & D program. Our report is not based on pairs—there are cases of more than one failed R & D program corresponding to a particular innovation, and in several cases there is no successful program at all.

2. A SAPPHO pair by definition belongs to a single market or area although the technological solutions of each program of the pair and the firm characteristics may both differ. In our study the areas or markets associated with the failed programs are not necessarily related to the areas or markets of the successful ones.

3. In SAPPHO a search is made for variables which differentiate success from failure (i.e., whose level is higher, or lower, in the successful member of the pair than in the failed member, for a significant number of innovations). These variables do not include market or area characteristics (see the section on choice of variables). In our report we search for variables which are correlated with performance, and they may include area or market characteristics.[2]

4. It follows that our methodology is more appropriate than SAPPHO's for studying a country's comparative advantage in different areas or markets of science-based industry. SAPPHO's methodology is, on the other hand, more adequate than ours for analyzing the relative importance for performance of other variables related to firm behavior.

---

2. For example, "rate of growth of the market" will be positively associated with performance if the proportion of successful programs, say, in the 5%–10% growth range is lower than the proportion of successful programs in the 10%–15% growth range.

### Methodology

The methodological issues of the paper refer to problems associated with the unit of analysis, the measurement of performance, and the choice of variables, each of which is discussed separately.

#### Unit of analysis

We term our unit of analysis "R & D program." An R & D program is a process or chain of activities ranging from the formulation of the original idea for a new product or products up to the marketing of this new product throughout its lifespan. It therefore comprises inputs (e.g., R & D, market surveys), output—the new product, and performance—the relation between the real value of the output and the real value of the inputs. Dividing the stream of innovative activity of a particular firm into separate R & D programs is useful if it provides a better appreciation of the factors explaining overall performance. In this context, an ideal R & D program should fulfill the following conditions: (1) that it is possible to measure performance precisely; (2) that it is possible to determine its characteristics unambiguously. The first of these is guaranteed by two conditions: (1a) that the program's outputs depend on its own inputs and not on the inputs of other programs; and (1b) that the program's inputs affect only its own outputs.

In the real world it is unlikely that a division of a firm's innovative stream into real R & D programs can be found. Any division is bound to involve complementarity (i.e., synergy) between the resulting R & D programs; that is, there exist R & D programs whose inputs or outputs affect the performance of other R & D programs. It is particularly useful to consider two kinds of synergy, in R & D and in marketing.

The first, synergy in R & D, occurs when the technical knowledge generated in a particular R & D program represents a useful input into one or more other R & D programs. This follows from the public-good nature of information; i.e., a piece of information never gets used up no matter how many times it is used.

Synergy in marketing occurs when the value of sales or marketing costs of the products of a particular R & D program are influenced by the fact that the firm is offering and selling products of other R & D programs. This may be due to better utilization of marketing facilities or when customers have a preference for suppliers offering a wide range of products over suppliers offering a narrow range.[3]

The almost inevitable synergy between a firm's various R & D programs does not enable performance in any one program to be explained exclusively in terms of that program's characteristics; the characteristics of other programs

---

3. For a more complete description of synergy see Ansoff 1968, chap. 5.

also have an influence. This does not mean that it is not useful to divide a firm's innovative stream into parts for separate analysis. It only tells us that the more synergy there is, the less such a procedure will contribute to our understanding of overall performance and that the practical issue is to find a set of programs with as little as possible synergy between them.

The possibilities of dividing the innovative activity of a firm into R & D programs are also limited by the level of aggregation of the available costs and benefits data. Our division was essentially determined by the files of R & D projects submitted to the Ministry of Commerce and Industry. The impossibility of distributing the R & D costs of a particular file among the various products in the file precludes our generating more R & D programs than files. However, when market or R & D synergy was sufficiently strong, we defined single R & D programs from more than one file.[4]

### Measurement of performance

We distinguish between two levels of performance, success and failure. The success-failure dichotomy is intended to differentiate between R & D programs which led to commercially successful new products and those which did not. The index of commercial success is given in most cases, for lack of more complete information, by the ratio of the real discounted value of sales to the real value of R & D costs.[5] Even with complete information on direct costs and benefits, there are at least two conceptual problems of measurement: synergy (in R & D and marketing) and the length of the period of observation. The latter is the reason for having a separate group of indeterminate programs.

*Synergy.* The problem of synergy is particularly difficult when the knowledge generated in a failed program has contributed significantly to a successful one. It is practically impossible to assign a shadow price to such a piece of knowledge. A related problem may arise in marketing where in order to capture a share of the market that ensures profitable operation a firm may have to launch a series of products belonging to several R & D programs. One or more of these programs may be failures in the sense that their direct profitability (obtained by comparing sales with R & D costs) is low or negative.[6]

---

4. This was done in two instances: an R & D program combining projects for successive generations of an instrument and one combining projects for an instrument and its peripherals.

5. Mansfield et al. (1971) show for a sample of chemical, electronic, and mechanical innovations that about half of the costs of *launching* a new product are R & D costs. The ranking of programs according to their R & D/sales ratio will coincide with their ranking according to the ratio of total cost (both at and after launch) to sales if the R & D/total cost ratio does not vary significantly from program to program.

6. In an extreme case of this kind one may be compelled to consider the set of projects corresponding to the whole line of products as a single R & D program.

*Length of the period of observation.* An R & D program is a series of activities which take place over time, and the question arises, at what moment of time is it justified to make an observation with the object of determining success or failure? In other terms, what are the criteria that tell whether an R & D program has been completed or what are the criteria for evaluating an uncompleted program? There is probably no unique answer: factors such as product life, time required for market penetration and acceptance, and even the firm's future plans for the program all would certainly influence the answer. This problem does not appear for programs suspended during development but it is very real for those whose products have been launched and whose sales show a continuous increase but for which no reasonable sales/R & D ratio has been attained at the moment of observation. Of course, it is possible to make predictions on the basis of past tendencies and so forth, but it is doubtful whether these could be combined with the data to produce a numerical indicator of success or failure.[7]

*Indeterminate programs.* Of the 35 R & D programs for which information was collected, 12 were defined as indeterminate. These are programs which at the time of observation were in one of the following situations: (*a*) last stages of development; (*b*) first stages of marketing; (*c*) later stages of marketing, but with no possibility of determining yet whether or not the program is successful. The indeterminacy of performance of these programs forced us to exclude them from the main body of the analysis. Our intention is to predict their performance in the light of the conclusions obtained from this report.

*Types of failure.* There are two types of failure within the R & D programs studied: suspension before market launch and suspension after market launch. There is no clear case of an *ongoing* R & D program which reached the commercial stage being classified under failed programs.[8] These categories are purely descriptive—we have not attempted in this paper to classify failed programs according to the cause of failure: technological, market, or firm. This will be done at a later stage.

### Choice of variables

The report identifies some variables which are correlated with the performance of R & D programs. They are a small subgroup of a larger group of variables which were initially (and in the early stages of the research) considered of possible relevance. We call the latter group the initial set of variables.

---

7. The problem is particularly difficult with completely new products, for which predictions of market penetration and product life are particularly uncertain. At any rate, such predictions would be based on the *results* so that it is not legitimate to include them as part of the data *input*.

8. Of course, many such programs were classified as indeterminate.

Table 1.1. Size of firms: 1972/73

| Number of firms | Turnover in thousands of I £ | Number of employees |
|---|---|---|
| 5 | up to 2,000 | up to 70 |
| 1 | 9,000 | 90 |
| 2 | 20,000–30,000 | 360–350 |
| 1 | 100,000 | 1,600 |

Table 1.2. Distribution of programs by field of medicine and function

| Field of medicine | Function | | | | |
|---|---|---|---|---|---|
| | Diagnostic[a] | Therapeutic | Monitoring | Lab. | Total |
| Cardiology | 1 | 3 | 4 | — | 8 |
| Pulmonary function | — | 1 | 1 | — | 2 |
| Nervous system | 2 | 3 | — | — | 5 |
| Urology | — | — | 1 | — | 1 |
| Dentistry | — | 1 | — | — | 1 |
| Nuclear medicine | 2 | — | — | — | 2 |
| Speech aids | — | 1 | — | — | 1 |
| Other | — | — | 2 | 1 | 3 |
| Total | 5 | 9 | 8 | 1 | 23 |

[a]Does not include diagnostic monitoring.

importance in analyzing their performance and imposes an additional constraint on the generality of our conclusions.[11]

*Size.* It is clear from table 1.1 that most of the firms are very small, even by Israeli standards. This has far-reaching implications for the degree of flexibility enjoyed by them and consequently for their strategy of choice of R & D programs

*Percentage of turnover devoted to R & D.* During 1972/73, five of the firms devoted 10%–25% of their turnover to R & D, two firms 30%–40%, one firm 50%, and one firm 70%, the average being 30%. This is a very high figure compared with foreign firms operating in the same field but is consistent with the fact that most of the firms are new. Indeed, there is a very strong negative correlation between age of firm and percentage of turnover devoted to R & D.

*Degree of specialization.* Only two firms specialized exclusively in biomedical

11. Correspondingly, the behavior of these firms may, in principle, reflect conditions resembling those of infant industries.

Table 1.3. Successful and failed programs

| | Number of programs | R & D costs | |
|---|---|---|---|
| | | Thousands of 1970 I £ | Percentage |
| Successful programs | 5 | 2,405 | 48.1 |
| Failed programs | 18 | 2,615 | 51.9 |
| Suspended before market launch | 10 | 1,985 | 37.8 |
| Suspended after market launch | 8 | 630 | 14.1 |
| Total | 23 | 5,020 | 100.0 |

[a]Nominal R & D costs converted to 1970 prices and cumulated over the period during which the program was carried out. The deflator used is the index of wages and salaries per employee in "electrical and electronic equipment" (Central Bureau of Statistics, *Statistical Abstract of Israel 1973*, no. 24, p. 439, table XIV/9; and earlier issues of the *Abstract*). This index was used because 60%–80% of R & D costs are salaries.

electronics; in another four it was a major field of activity; and in the remaining three, biomedical electronics was only a minor field.

We gathered information about 37 R & D programs undertaken by the 9 firms up to 1974. Two of them were dropped because not enough information was available. Another 12 were classified as indeterminate, and the analysis was carried out on the remaining 23. The basic information on these 23 programs is summarized in tables 1.2 and 1.3

### Analysis: Origin of Idea

An attempt was made to classify each R & D program according to whether the basic idea leading to it did or did not originate in R & D. This distinction refers both to the segment of the industry in which the idea originated and to the type of perception involved. Thus ideas originating in R & D originate from R & D personnel and involve the perception of a technological opportunity. If the idea does not originate in R & D, then it originates from users, agents, salesmen, or a product manager (who in well-established firms lies at the interface between technology and the market); and the perception refers to a *specific* need or to a combination of need and technical opportunity which cannot, conceptually or statistically, be separated from each other.[12]

12. It is worth pointing out that the idea may originate within the firm or outside it. For example, program ideas originating in R & D within the firm may represent a spin-off from other R & D programs or be the result of a search for applications of a component previously developed by the firm or its manager; ideas originating in R & D but coming from outside the firm may involve outside researchers or laboratories providing a research idea, research results, or a prototype for further development and commercial exploitation.

Table 1.4. Success and failure of programs, by origin of idea

|  | Origin of idea | | |
|---|---|---|---|
|  | R & D | Other | Total |
| Failure | 12 | 1 | 13 |
| Success | — | 4 | 4 |
| Total | 12 | 5 | 17 |

*Note*: Six of the programs could not be classified owing to lack of information.

The distinction of successes and failures among the origin-of-idea categories mentioned above may tell us something about the pattern of linkage of technology and market most conducive to success in the Israeli biomedical electronics sector—in particular, whether an optimum product-definition strategy should start with a search for technological opportunity or with a search for a market or need. More generally, it may also provide an indication of the relative importance of technological and market factors in ensuring success. At any rate it should be evident that no clear-cut conclusions can emerge from an analysis of the origin of a program idea without explicitly considering other variables more directly related to technological and market risk.

Table 1.4 shows the distribution of successes and failures according as the idea originated exclusively in R & D or not, and indicates that *the proportion of failures in programs whose idea originated in R & D exceeds the proportion of failures in the other programs.* Success in product innovations within a particular sector is the result of, among other things, an adaptation of technology to market requirements that is optimum for the sector. The process may be viewed as beginning with the program idea, whose origin may determine its future development and final outcome. The empirical findings support the view that there is a need-perception component at the origin of an R & D program which has achieved an optimum link between needs and technology.[13] In this context, the large number of failures reflects the fact that most of the firms were either new or recent entrants to the sector. These firms have to get hold of one or more research ideas with which to begin operating. They seldom have other alternatives since these can only result from a network of market feedback, which takes a relatively long time to build up.[14]

13. The results obtained here should be viewed as complementary to the results obtained later on in our analysis of market determinateness.

14. Well-established firms, on the other hand, have access to program ideas originating from R & D and elsewhere and are also in a much better position to adapt technology to market needs. Thus, even programs originating in R & D may succeed.

## Analysis: Market Determinateness

A central variable in our analysis of performance in innovation is the degree of market determinateness of the products of R & D programs.[15] Market determinateness refers to the degree of specificity of the market signals received by the innovating firm and consequently to the extent to which it anticipates (instead of responding to) demand.

In order to explain the concept we introduce four types of market signal, in ascending order of specificity: (1) signals about a need, (2) signals about a product class, (3) signals about basic functions, (4) signals about product specifications.[16]

Suppose that a firm launches a product in a known market. The product then represents a response to signals about a need and a product class (specificity level 2). Whether more specific signals are also present depends on what kind of market is being considered: if the market is sufficiently mature that the functions of the product have already been defined (standard functions), then clear signals about the functions are also present. In this case, barring an R & D project commissioned by the final user, the firm anticipates demand with respect to specifications only. If, on the other hand, the basic functions have not been standardized, then no clear signals about them have been received by the firm. It therefore anticipates demand both with respect to specifications and with respect to basic functions.[17] Table 1.5 summarizes the signals received and the demand anticipation implied by the five possible categories of market determinateness of R & D programs.[18]

We postulate, ceteris paribus, a positive link between a program's degree of market determinateness and the expected level of (*a*) the product's *objective capacity* to satisfy user needs, (*b*) *recognition* of this capacity on the part of the

15. This section benefited greatly from an illuminating discussion with A. Suhami at an early stage of the work.

16. In this context signals are about product characteristics or needs and not market prices (or quantities).

17. Let us try to clarify what are *basic functions* and what are *product specifications* in the context of biomedical electronics. For example, each one of the physiological parameters measured in a cardiovascular monitoring system (ECG, blood pressure, etc.) would represent a basic function. Product specifications on the other hand would determine the efficiency, reliability, simplicity of operation, and degree of patient safety with which the basic functions are performed. These are in part related to the data-processing and display capabilities of the system. Synchronous defibrillation is a *standard* basic function in the treatment of ventricular fibrillation and other heart ailments. A fuller discussion of standard versus nonstandard basic functions in biomedical electronics follows at the end of this section.

18. The proposed categories of market determinateness are to some extent arbitrary. For example, product specification may become increasingly standardized with the passage of time. If this happens, another category of market determinateness between levels 4 and 5 may be warranted.

Table 1.5. Specificity of signals and market determinateness

| Degree of market determinateness | Type of signal ($\sqrt{}$) or type of demand anticipation ($\times$) | | | |
| --- | --- | --- | --- | --- |
| | About need (1) | About product class (2) | About basic functions (3) | About product specification (4) |
| 1. No clearly perceived need | $\times$ | $\times$ | $\times$ | $\times$ |
| 2. Clearly perceived need but product unknown | $\sqrt{}$ | $\times$ | $\times$ | $\times$ |
| 3. Product known, but basic function not completely standardized | $\sqrt{}$ | $\sqrt{}$ | $\times$ | $\times$ |
| 4. Product known and basic function standardized | $\sqrt{}$ | $\sqrt{}$ | $\sqrt{}$ | $\times$ |
| 5. Commissioned R & D program | $\sqrt{}$ | $\sqrt{}$ | $\sqrt{}$ | $\sqrt{}$ |

user, and ($c$) his *willingness* to consider the product when a purchasing decision is to be made.[19]

The reason for this is the negative relationship between market determinateness and the degree of uncertainty both on the part of the innovator firm and on the part of potential users. The less specific are the signals, the greater the uncertainty of the innovating firm about the precise composition of user needs (and the less defined these needs are), and the more difficult potential users find it to evaluate the product's objective capacity to satisfy their needs (and again the less defined these needs are). It might be useful to separate two aspects of recognition: identification of product ($b_1$) and evaluation of its concrete usefulness ($b_2$). Identification of a product simply means assigning it to a product class linked with the satisfaction of a particular need, i.e., awareness of its potential usefulness in a very general sense. The problem of identification exists when the R & D program falls in the low or medium market-determinateness categories. The second aspect, $b_2$, concerns evaluating the

19. The degree of market determinateness is also presumably negatively related to the variance ($a$)–($c$). In what follows we consider individual products separately and ignore for the time being possible spectrum effects on ($c$); i.e., the fact that a *line* of products is offered by the firm may have a positive effect on the willingness of potential users to consider an individual product when a purchasing decision is made. These effects—which may be significant when market determinateness is high—will be considered at a later stage.

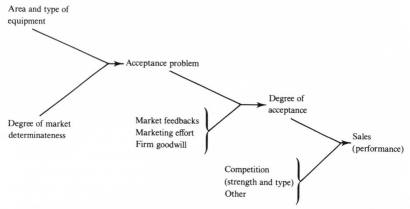

Figure 1.1. Market determinateness and performance. The category labeled "Other" includes, e.g., program attributes such as innovation profile, area variables such as market size and growth, and firm characteristics such as firm size.

concrete usefulness of the product. The user's ability to do so increases as previously undefined functions of the product gradually become standardized.

Several remarks arise in this connection. (The discussion which follows is summarized schematically in fig. 1.1.) First, (a)–(c) will be affected not only by the degree of market determinateness—which is an area characteristic—but also by other factors pertaining to both the firm and its strategy. We have selected three: ($\alpha$) the efficiency of the firm's market feedback (firm and strategy), ($\chi$) marketing effort (strategy), and ($\gamma$) the firm's goodwill. Each of these factors would primarily affect the corresponding variable in set $a$–$c$; e.g., the more efficient the market feedback enjoyed by the firm, the greater, ceteris paribus, the product's objective capacity to satisfy user needs. Indirectly each factor affects all three variables Second, each variable is related to the next; thus the greater the product's objective capacity to satisfy user needs, the greater (in general, presumably, though certainly not always) the recognition of capacity and the willingness to consider the product.[20]

Product acceptance is a measure of the willingness of users to consider the product when a purchasing decision is to be made (variable $c$). It is a necessary condition for a product to be demanded, but, for at least two reasons, not a sufficient one: the timing of a user's purchasing decision may be independent

20. In general, the discrepancy between ($b$) and ($c$) is due to one or both of, first, spectrum effects (mentioned above) and, second, the extent to which users are willing to purchase from a firm despite the uncertainty surrounding the objective capacity of the product to satisfy their needs. Thus ($c$) will depend in part on the firm's goodwill.

of the acceptance times of the various alternatives; and a particular accepted product has to compete with other accepted products. Thus, acceptance and competition jointly determine the demand for a product.[21]

We can now specify the two variables or groups of variables determining product acceptance: first, the degree of market determinateness and, second, the variables $\alpha-\gamma$ (extent of market feedback, marketing effort, and the firm's goodwill). The degree of market determinateness determines the magnitude of what can be termed the acceptance problem, which has to be overcome for success, and, for given levels of variables $\alpha-\gamma$, the risk of not gaining enough acceptance fast enough to ensure success (acceptance risk).[22] Variables $\alpha-\gamma$, to which may be added the passage of time, determine the extent to which the acceptance problem has been overcome. Ceteris paribus, the greater the acceptance problem, the greater the marketing effort required for success.

The seriousness of the acceptance problem (even within a given category of market determinateness) is a function of other parameters as well, such as area, type of instrumentation, and the characteristics of users.

In some areas the existence of a clearly perceived need is all that is required for immediate acceptance. For example, a completely new drug whose purpose is to cure acute cases of cancer may be tried and accepted almost immediately. The lack of complete information on its effects may not be so relevant for treatment of critical cases. Alternatively, a new pacemaker, even if completely standardized, may involve a serious acceptance problem: lack of minimum information may have serious consequences for the ultimate user. These extreme examples suggest that the acceptance problem of strongly determined products in some areas may be greater than the acceptance problem of weakly determined products in other areas.[23] In biomedical electronics, the acceptance problem may be less serious for diagnostic equipment (in any category of market determinateness) than for monitoring or therapeutical equipment. We conclude that the correlation between the degree of market determinateness and the magnitude of the acceptance problem may hold only within a given area.[24] Similar considerations would hold for the type of equipment developed and the characteristics of users in any given area. Thus the more

21. Partial acceptance of new biomedical products is gained after successful tests by recognized specialists in the field and publication of the results.

22. Acceptance risk should be distinguished from another market risk deriving from competition.

23. Within each of these extreme areas, our market determinateness categories would probably not reflect strongly varying magnitudes of the acceptance problem and would presumably not differentiate success from failure.

24. In a wider framework, the distinction between areas involving final consumer goods and those involving producer goods may also be relevant.

Table 1.6. Success and failure, by market-determinateness category

| | | Market-determinateness category | | |
|---|---|---|---|---|
| | Total | High | Medium | Low |
| Number of programs | | | | |
| Failure | 18 | 4[a] | 7 | 7 |
| Success | 5 | 5 | — | — |
| Total | 23 | 9 | 7 | 7 |
| Percent of total R & D costs | | | | |
| Failure | 51.9 | 32.0[a] | 10.2 | 9.7 |
| Success | 48.1 | 48.1 | — | — |
| Total | 100.0 | 80.1 | 10.2 | 9.7 |

[a] Additional information may lead us to reclassify two R & D programs involving 25.8% of total R & D costs into the medium-market-determinateness category. In this case, our conclusions will be reinforced.

complex the equipment and the greater the number of users, the greater the acceptance problem.

### The findings

Each R & D program was allotted to one of three market-determinateness categories: *low* (or *weak*), unknown products (with or without a clearly defined need; corresponding to categories 1 and 2 of table 1.5); *medium*, known products whose function is not fully standardized (category 3 in table 1.5); and *high* (or *strong*), known products with standardized basic function (categories 4 and 5 in table 1.5). The category low market determinateness can be subdivided according to whether (*a*) the specific function of the new product was being performed before by other means, either more imperfectly or more expensively, or both; or (*b*) the specific function of the new product was not being performed by other means before the appearance of the product. The category medium market determinateness is subdivided into (*a*) products designed for experimental purposes which have not yet reached the stage of standard clinical application and (*b*) products which perform a basic function which is already standard in some clinical applications but in a radically different way. In this case the product may be suitable for existing applications or it may create new applications or both. High market determinateness includes products whose basic functions are standard in some clinical application.

Two comments on these categories are in order. In the category of low market determinateness, we can be sure that the products satisfy a *specific* need only in the case of (*a*). The products of R & D programs in (*b*) probably do not satisfy a well-perceived specific need but may still satisfy a more general or a latent need. In the category of medium market determinateness, the issue arises as to when the introduction of a radically new way of operation makes the basic functions of an instrument less standard than they are according to the conventional method or working principle.[25] The distribution of successes and failures among the three market-determinateness categories can be observed in table 1.6.

Several conclusions can be drawn from the table. First, the proportion of successes drops to zero as we go from the high-market-determinateness category to medium and low. Second, the number of failures increases as we go from high to medium or low. Third, most programs with high market determinateness are successes, and most of the R & D spent in this category is associated with success. We may thus state that the degree of market determinateness differentiates sharply between the successful and the unsuccessful biomedical electronics R & D programs studied. There is no need, in this connection, to differentiate areas and types of instrumentation.[26]

### Analysis: The Innovation Profile

The innovation profile underlying a particular R & D program involves both the technical (new product) component and the price of the program's product, both in relation to products already in the market. In this study we distinguish three innovation categories referring to the technical component:

*A.* Completely new instruments.

*B.* Similar instruments exist, but the product of the R & D program seeks to perform the basic functions of the instrument in a radically different way, a way which is new in the existing fields of application.

*C.* The products of the R & D program perform the basic functions of existing instruments in basically the same way[27] although usually more efficiently.[28]

---

25. In a more general framework, the criteria for deciding whether or not the introduction of new technology into existing products reduces their market determinateness should take account of the extent to which the new technology is visible to the user and the extent to which the new technology requires adaptability on the part of the user. These factors were mentioned to us by D. Rosenbloom. See also n. 29 below.

26. This result is consistent with the findings of Jones (1969, 18–22) concerning what he terms the leading problems facing a sample of 265 biomedical firms. "Determining product requirements of the market" and "achieving market acceptance" were considered to be leading problems by the greatest and third-greatest number of firms, respectively (Jones 1969, table 15).

The result also fits in with Hirsch's (1967, chap. 3) qualification to his main product-cycle model result which states that small developed countries have a comparative advantage in new products. The range of new products for which this is true narrows when export marketing (including feedback and the collection of market information) is much more expensive than domestic marketing.

27. The introduction of new components into existing instrumentation often fits in here.

28. E.g., with higher data-processing and display capability, with increased patient safety and reliability, and more simply and comfortably.

Table 1.7. Distribution of programs and R & D costs by degree
of market determinateness and innovation category

| Degree of market determinateness | Number of programs in innovation category | | | R & D share (%) in innovation category | | |
|---|---|---|---|---|---|---|
| | A | B | C | A | B | C |
| Low | 7 | — | — | 9.7 | — | — |
| Medium | — | 4 | 3 | — | 6.4 | 3.8 |
| High | — | 2 | 7 | — | 25.8 | 54.3 |
| Total | 7 | 6 | 10 | 9.7 | 32.2 | 58.1 |

Table 1.8. Performance by innovation category

| | Total | Innovation category | | |
|---|---|---|---|---|
| | | A | B | C |
| Number of programs | | | | |
| Failure | 18 | 7 | 6 | 5 |
| Success | 5 | — | — | 5 |
| Total | 23 | 7 | 6 | 10 |
| R & D share (%) | | | | |
| Failure | 51.9 | 9.7 | 32.2 | 10.0 |
| Success | 48.1 | — | — | 48.1 |
| Total | 100.0 | 9.7 | 32.3 | 58.1 |

The innovation categories are related to the market-determinateness categories, although—with the exception of category $A$, which is equivalent to the
low-market-determinateness category—they are not identical with them. Thus
category $B$ includes R & D programs with both medium and high market
determinateness; i.e., a radically new way of performing an existing function
may or may not change the standard functions and applications of the instrument.[29] In addition, R & D programs in category $C$ need not have high market
determinateness if the basic function-applications relation has not yet been

29. A radically new way of performing an existing function is likely to change a product's
efficiency in various directions. If efficiency in existing applications increases, the new product will
be considered strongly market determinate if the existing function and applications were already
standardized. (In this case, there will be a widening of the field of application of the function). If
efficiency increases in one direction and decreases in another, existing standards in function applications are not relevant in regard to the new product and new standards should be created. In this
case the program is only of medium market determinateness and there may be a decline in the
instrument's range of application.

standardized. The relation between innovation category and market determinateness is illustrated in table 1.7.[30]

The distribution of success and failure by innovative category is shown in table 1.8. We can see that all the successes belong to category $C$. In addition, they also belong to the high-market-determinateness category in table 1.7. We can thus say that a necessary (but not sufficient) condition for an R & D program to be successful was that the products of the program perform the central functions of existing instruments in basically the same way, usually more efficiently (category $C$).[31] In addition, a higher share of the R & D costs of programs satisfying this condition was associated with success than with failure. This conclusion can be interpreted as follows: R & D programs belonging to category $C$ that are only medium market determinate are likely to be market failures, while strongly market-determinate projects belonging to category $B$ may fail technologically.[32]

### Price comparisons

We have referred to the technical side of the product innovations covered in this report while almost completely neglecting their price. A complete characterization of an innovation, i.e., the innovation profile, must also consider this element and relate it to existing competing products. At least two aspects of price seem relevant, a comparison of absolute prices and a comparison of prices per efficiency unit in the performance of basic functions.[33] The relative importance of the latter seems to increase as basic functions and their applications gradually become standardized. When a particular basic function/application has yet to be standardized—when objective yardsticks for performance in specific applications have not yet been established—it may be more problematic to translate an increase in function efficiency into medical terms.

### Alternative profiles

Given its objectives and constraints, a firm should decide on the innovation profile which will maximize its chances of success in the R & D programs or program it intends to undertake. We attempt to relate innovation profile and

30. It can be seen that two programs in category $B$ representing 25.8% of total R & D have strong market determinateness, while three programs (representing 3.8% of total R & D) in category $C$ have medium market determinateness.

31. A large proportion of R & D programs in categories $A$ and $B$ were undertaken under considerable technological uncertainty. All except two of the programs were either weak or medium market determinate. It would seem that a high degree of risk, either market or technological or both, accompanied these two categories.

32. A more rigorous analysis of failure goes beyond the scope of the present paper.

33. By definition price comparisons cannot be made for programs belonging to category $A$.

Table 1.9. Performance of strongly market-determinate programs in innovation category $C$

|  | Total | $C_1$ | $C_2$ |
|---|---|---|---|
| Number of programs |  |  |  |
| Failure | 2 | — | 2 |
| Success | 5 | 4 | 1 |
| Total | 7 | 4 | 3 |
| R & D share (%) |  |  |  |
| Failure | 6.2 | — | 6.2 |
| Success | 48.1 | 46.9 | 1.2 |
| Total | 54.3 | 46.9 | 7.4 |

performance only for category $C$ since it is only here that we find both success-ful and failed R & D programs. For the same reason we consider only those programs which are highly market determined (table 1.6). This leaves us with a set of seven programs covering 54.3% of the investment in R & D.

All the programs in the set showed an increase in function efficiency per unit of cost, but the question we ask is whether the main thrust of the innova-tion lies in increased technical efficiency or in decreased cost. For this purpose we divide these programs into two profile subgroups:

$C_1$. Significant improvements in function efficiency with a price which is either similar to or above the price of competing instruments.

$C_2$. The new instruments might include improvements as in $C_1$ but they are less significant, the emphasis being on reducing the price of the instrument compared with competing products.

The distribution of success and failure between $C_1$ and $C_2$ is shown in table 1.9, from which we infer that all highly market-determinate programs in category $C$ involving a significant improvement in function efficiency and sell-ing at an equal or higher price (i.e., profile $C_1$) succeeded. Two out of the three highly market-determinate programs (and most of the R & D expenditure) in category $C$ whose main thrust lies in price reductions failed.

## Type of competition

A possible explanation for the positive relationship between performance and the emphasis on increasing function efficiency (in contrast to, and even at the expense of, reducing the price) lies in the type of competition in the biomedical market. Competition is mainly based on subjective confidence in supplier and quality, and not (predominantly) on price.

Knowledge of the relative importance of quality and servicing efficiency versus price and quality and servicing efficiency versus subjective confidence in supplier may enable firms to plan their strategy more efficiently. Six of the nine firms explicitly ranked the relative importance of price, quality, servicing efficiency, and subjective confidence in supplier. Two gave different rankings for each of two projects, and in one case the first place was given to two factors.

We summarize the findings briefly. (1) Subjective confidence in supplier is ranked as the dominant form of competition (and consequently as the most important factor for success) in most cases (five out of nine). (2) In the remaining cases, quality and servicing efficiency (the same number of cases for each) were referred to as the dominant form of competition. (3) In no case was price considered the dominant form of competition.

Despite the small number of observations and the lack of direct evidence from users on the dominant forms of competition, the results coincide with those obtained in a recent study.[34] They suggest several interesting points. The importance of subjective confidence in supplier is a direct result of users' lack of complete information about the performance characteristics of the products purchased.[35] This is why a firm's goodwill affects the degree of acceptance achieved (see fig. 1.1). Recent entrants into the biomedical field will very likely not enjoy the confidence of their potential customers: their competitive edge must be in the possibility of supplying higher quality products.

In general a small unknown entrant should, in view of the forms of competition dominant in the industry, begin by offering goods whose objective quality is high and easily perceptible, and whose servicing requirements are demonstrably small. Easily perceptible quality and small servicing requirements reduce the uncertainty due to lack of complete information and they therefore reduce the importance of subjective confidence in supplier (the factor which is not available to new firms). Instrumentation which satisfies these two conditions is characterized by two features: it is simple (and small) and its functions are standard (making easy evaluation of performance possible).[36] If customers' expectations concerning the first products are confirmed, subsequent sales will enjoy greater confidence, which might enable the firm to increase the size and

34. Peterson and MacPhee (1973) have emphasized the dominance in the purchasing decisions of 246 medical-care professionals of "nonperceptible" differences in products (such as firm's reputation and recommendation of colleagues) over "perceptible" differences (such as quality and price). Moreover, the various aspects of quality (specifications, safety, ease of operation, etc.) were all more important than price in determining purchasing decisions.

35. The lack of information refers to both quality and servicing, but it is probably more acute for servicing. This is because at least some aspects of quality can be perceived at the time of purchase. But servicing is only provided in the future, so its objective evaluation in the present is difficult.

36. These conclusions again complement those reached in connection with market determinateness (with the additional point that a simple reduction in price will not work).

complexity of the systems developed. Because of the general acceptance problem discussed above, development of products incorporating unstandardized functions is even then not recommended.

The implied relatively low elasticity of demand with respect to price merits careful analysis. Here, however, we can only sketch some of the implications. Lack of search time and inadequate information on alternative products do not permit users to consider simultaneously all the quality-price features of all alternatives. The result is a process of choice and rejection by stages, starting with quality or confidence in supplier and only then proceeding to price comparisons. Much more remains to be done in conceptualizing this kind of purchasing decision.

### Conclusions

1. The report has isolated a reduced set of variable which are correlated with the innovation performance of firms belonging to the Israeli biomedical electronics sector. The central variables are market determinateness (an area characteristic) and the innovation profile (an attribute of R & D programs). The nature of the idea leading to R & D programs, the type of competition prevailing in the industry, and the notion of product acceptance all complement the main explanation given by these two variables. Consideration of additional program, firm, and area variables—such as the spectrum of products offered by the firm, market and firm size, and rate of growth—will undoubtedly contribute further to our understanding of performance in innovation. This task is left for a later stage.

2. Since the objective access to markets and market information does not seem to have differed much between firms,[37] it follows that differences in performance at the R & D program level basically reflect differences in management, i.e., in the extent to which the variables mentioned above were taken into account in the strategic decisions of the firm.

3. Although the variables selected for the study managed to explain performance of past biomedical electronic R & D programs, we should be cautious in using them to predict performance of future programs without taking into account changes in the characteristics of the firms involved. For most of the programs studied, the firms were small or newcomers to the field or scientific entrepreneurs, or all three. Their size has now changed and also their approach and strategy—and in some cases the time may be ripe for taking a calculated risk in a promising new area despite its not being very market determined.

---

37. For example, when comparing the performance of American and European firms in science-based industry, an OECD study argues that the relatively greater access to the United States market of firms in the United States has been a significant contributing factor. See Organization for Economic Cooperation and Development 1968.

### References

Ansoff, H. I. 1968. *Corporate Strategy*. Harmondsworth: Penguin Books.

Centre for the Study of Industrial Innovation. 1971. *On the Shelf: A Survey of Industrial R & D Projects Abandoned for Non-Technical Reasons*. London.

Freeman, C. 1965. "Research and Development in Electronic Capital Goods." *National Institute Economic Review* 34 (November): 40–91.

Hirsch, S. 1967. *Location of Industry and International Competitiveness*. Oxford: Clarendon Press.

Jones, H. R. 1969. *Constraints to the Development and Marketing of Medical Electronic Equipment*. Ann Arbor: Institute of Science and Technology, University of Michigan.

Mansfield, E., et al. 1971. *Research and Innovation in the Modern Corporation*. New York: Norton.

Myers, S., and D. G. Marquis. 1969. *Successful Industrial Innovations*. National Science Foundation, NSF 69-17. Washington, D.C.

Organization for Economic Cooperation and Development. 1968. *Gaps in Technology: Electronic Components*. Gaps in Technology between Member Countries Series. Paris. (See also other volumes in this series.)

Peterson, R. D., and C. R. MacPhee. 1973. *Economic Organization in Medical Equipment and Supply*. Lexington, Mass.: Lexington Books.

Science Policy Research Unit, University of Sussex. 1972 *Success and Failure in Industrial Innovation: Report on Project SAPPHO*. London: Centre for the Study of Industrial Innovation.

This report presents the initial results of a research project on performance in innovation in the Israeli electronics industry directed by Morris Teubal. The original proposal, entitled "Threshold R & D Levels in the Israeli Electronics Industry," was jointly prepared with Naftali Arnon. Manuel Trachtenberg has since joined the group and actively contributed to the design of the study and the writing of the report. The project is being carried out at the Maurice Falk Institute for Economic Research in Israel, Jerusalem, with the financial participation of the National Council for Research and Development, the Prime Minister's Office, Jerusalem.

We especially appreciate the cooperation of Y. Amir, M. Barone, M. Ben-Porat, S. Ishai, Yona Mahler, U. Peer, Z. Rosner, Z. Shalev, M. Silberman, A. Suhami, A. Wilenski, and R. Zernik. We have also benefited from discussions with M. Eshel and Z. Zelinger and the cooperation of the Ministry of Commerce and Industry, Office of the Chief Scientist. We are grateful to C. Freeman for his encouragement and to S. Kuznets for his valuable advice. T. Blumenthal, D. Caplin, K. L. R. Pavitt, R. Rosenbloom, R. Rothwell, and P. Spiller have made useful comments on an earlier draft of the paper. S. Freund has provided useful editorial help.

# 2 Analysis of R & D Failure

## Pablo T. Spiller and Morris Teubal

### Introduction

Previous work has revealed the explanatory power of the variable "market determinateness" in explaining performance in innovation in the Israeli biomedical electronics sector. This variable indicates the extent to which needs have been translated into product classes, functions, and specifications. For example, the need to reduce costs by 20% is a very general need with no expression in terms of products or specifications. When the user demands a particular type of cost-reducing device with certain characteristics, the need is very specific. In the first case, there is a high risk of failing to understand the precise structure of user needs and this structure may not be very well defined. Moreover, the product-acceptance risk derived from the user's incomplete information on the performance characteristics of the product is likely to be very high. These risks diminish when market determinateness increases. A formulation of this variable in terms of signals and the translation of the theoretical variable into empirical categories which suit biomedical instrumentation were achieved in Teubal, Arnon, and Trachtenberg (1976). The results show that the five successful programs (out of a total of 23 considered) all belonged to the strong-market-determinateness category where the basic function of the instrument is already standard in some clinical application. Programs belonging to the medium- and weak-market-determinateness categories (a total of 14) were all failures.

The basic issue we want to investigate in this paper is whether this performance pattern reflects the pattern of comparative advantage within the biomedical electronics sector for the period studied, given the collection of firms active in the sector,[1] or whether it also reflects inadequate execution of innova-

---

1. Comparative advantage here refers to the appropriateness of the innovations undertaken given the characteristics of the sector's firms and the conditions of the Israeli economy. Character-

tion programs or the effects of uncertainty.[2] For this purpose we take the group of 18 failed programs and we introduce three levels of inadequate firm behavior, corresponding to program execution, program selection, and overall firm behavior.

A second objective of the paper is to explain firm behavior in connection with the failed programs in terms of characteristics of the sector and of firms and to make explicit the way in which management conceives the innovation process which underlies inadequate search for extramural information, both before and after program selection.

### Framework of Analysis

The 18 failed innovation programs are analyzed under the following four headings: (1) stage of program suspension, (2) events leading to suspension, (3) evaluation of firm behavior, and (4) interpretation of firm behavior. Each of these points is here discussed briefly, the last three being dealt with in more detail in subsequent sections.

The failed programs are classified as suspended before completion of prototype (five programs), before market launch (five), and after market launch (eight). These are *descriptive categories* which tell us only at which stage of the innovation process the program was discontinued. There is no necessary correspondence between them and the explanatory categories commonly referred to in the literature as technological failure and market failure.

Events leading to suspension are the immediate reasons stated by the entrepreneur that have induced him to suspend the program. They have not in themselves led to failure but are rather a manifestation of such causes as inappropriate firm behavior or uncertainty. This distinction between the proximate reasons for failure and the causes of failure is of prime importance in the analysis of failure. The former also help us to evaluate and interpret firm behavior and the role of uncertainty. For example, program suspension due to belated realization of the marketing effort required is under certain conditions prima facie evidence of wrong program selection.[3]

Broadly speaking, failure may be caused by inappropriate firm behavior, by the workings of uncertainty, or by a combination of the two. It is of course difficult to distinguish between them, primarily because we do not know what

---

istics of management are not referred to when evaluating the appropriateness of innovations. A broader concept would have to consider these characteristics as well and in particular to recognize that the quality of management increases with the accumulated experience of the sector.

2. An alternative statement is that we are interested in studying the cause of failure while recognizing that the variable which most significantly differentiates success from failure across the universe of programs need not be the fundamental "cause" of failure in each program individually.

3. On this point see the section Events Leading to Program Suspension.

information firms had or could have searched for at the time. Consequently it is almost impossible to avoid contaminating our evaluation of firm behavior by hindsight; we hope that our results are nevertheless useful.

We distinguish three levels of inappropriate firm behavior: (1) overall firm level,[4] (2) choice of program, and (3) program execution. Inadequacy in program execution need not be the cause of failure when program selection was inappropriate (in fact it might be merely a reflection of the latter).[5] In other words, although inadequate behavior might be detected at both the program execution and program selection levels, the former represents a higher-level deficiency which overrides the latter. In contrast, it may be very difficult to evaluate the relative significance of inadequate firm behavior at both the program and portfolio selection levels when both are present simultaneously.[6]

Once inappropriate firm behavior has been detected there remains the question of explaining it. Broadly speaking the firm's behavior in connection with program selection and execution is governed both by the type and extent of search activity undertaken and by the firm's reaction to uncertainty. Both are influenced by firm characteristics such as the orientation of management, past experience, and present capabilities. The three main characteristics used throughout the paper as elements in the evaluation and interpretation of firm behavior are (1) orientation of management, (2) extent of involvement in the market, and (3) firm size.

We distinguish two types of management orientation. Technological orientation involves scientific or technological origin and lack of experience in management in general and marketing in particular.[7] Business-oriented management on the other hand necessarily implies previous experience and skills in management and marketing.

A technologically oriented management tends to underrate the importance

---

4. Of the various aspects of overall behavior, we concentrate here on the firm's choice of the set of innovation programs. We refer to this as portfolio selection.

5. In order to evaluate firm behavior at the overall and program selection levels we have to consider firm characteristics (such as size) and program or area characteristics (such as market determinateness and technological risk).

6. The answer may depend on the share of the firm's innovation resources taken up by the program. If it is high, the financial ability to pursue the program will probably not be very much affected by the misfortunes of other inadequately selected programs. Inadequate program selection will therefore probably be more significant in causing failure than inadequate portfolio selection. The contrary is probably true when the program's share of the firm's innovation resources is low. A more complete resolution of this issue will have to take into account, first, that appropriateness of program selection can be evaluated only in the light of a given portfolio and, second, that there is a time element which should be explicitly considered when dealing with portfolio selection.

7. It follows that scientific or technological entrepreneurs who have acquired marketing or management skills are not classified as technologically oriented.

of the marketing function (or product promotion) in innovation. In an extreme case it views the innovation process as consisting almost exclusively of efficiently executed R & D work.[8] This misunderstanding is inevitably reflected in the firm's objectives and strategy: there is a tendency for a technologically oriented management to specialize in a particular technology or scientific discipline (such as electronics) and not necessarily in particular markets or product lines. A business-oriented management would view the innovation process primarily as a process of linking technology with markets. R & D is then only one stage in the whole innovation process, and successful innovation implies overcoming other constraints as well.

The characteristic of market involvement should give us an idea of the firm's reputation among potential users, its knowledge of user needs, and its experience in marketing. These are very important factors in biomedical electronics and may well be critical for performance.

In the context of this paper, firm size (measured by turnover or employment) is best regarded as a measure of the firm's financial capabilities. It will thus determine which innovation programs are feasible and—via the cost of finance—which are profitable.

Four and possibly six of the seven firms considered here had technologically oriented managements in the period relevant to the analysis. Six out of the seven had no previous involvement in the markets corresponding to the programs studied. Lastly, at least four firms were small (70 or fewer employees).[9]

### Events Leading to Program Suspension

Program suspension is the result of a reevaluation of the possibilities of commercial success resulting from recent events or recently perceived program requirements. These we call events leading to suspension, or proximate reasons of failure. Their significance in an analysis of failure is twofold: first, they are distinct from the causes of failure, and, second, they provide clues to these causes. A classification of these events should take the latter into account in connection with, first, the relative importance of firm behavior and uncertainty in determining failure and, second, the significance of the various levels of firm behavior.

A classification relevant to the first point should take into account whether the events had occurred or the requirements existed when it was decided to embark on the program; if so, whether information on them was obtainable,

---

8. The conception which implicitly underlies this view is one of interchangeability between R & D effort and marketing. For further discussion see the section The Pattern of Decision Making.

9. Average employment in all Israeli manufacturing firms engaged in R & D was 184 in 1970/71.

and if not, whether they were predictable. In this paper we focus on the second point. In view of this we propose the following classification of proximate reasons for failure: (1) belated realization of past program requirements, (2) belated realization of future program requirements, (3) belated realization of low market potential, (4) difficulties in the execution of planned tasks, (5) purely financial, and (6) exogenous. Let us now discuss each of these event categories in turn and comment on their links with firm behavior.

1. This refers to additional aspects of the technology and markets of the innovation which the entrepreneur becomes aware of during program execution. Included here are belated awareness of promotion and marketing requirements (three instances), spectrum and line requirements (three), user needs (two),[10] clinical testing requirements (two), and R & D threshold (one), making a total of 11 in this category. The additional R & D testing and marketing requirements of the program which derive from these additional aspects (and which existed and were obtainable at the time of selection) lead the entrepreneur to discontinue execution (i.e., to cease to consider the program feasible or profitable). Events in this category almost inevitably point to inadequate firm behavior at the program execution level. Moreover, given their connection with program suspension, they also indicate program inappropriateness for the firm at the time of suspension and are an a priori indication of inappropriate program selection.

2. Belated awareness of future program requirements leading to program suspension suggests that inadequate behavior or uncertainty must have operated at the program selection level and not through any flaw in program execution, though the latter cannot be ruled out. Events in this category (four instances) were all concerned with belated awareness of promotion requirements.[11]

3. What has been said under 2. holds also for belated awareness of low market potential.[12] There were (two) instances in this category. As can be seen, in categories 1–3 events related to the market predominate over events related to technology.

4. Difficulties in the execution of planned tasks (five instances) involve technical problems in achieving the planned goals of development (or in testing prototypes) whose precise nature could not be foreseen. They may point to

---

10. User needs are defined in a narrow sense and do not include spectrum and line requirements.

11. In most instances in categories 1 and 2 the constraints or requirements have monetary implications, and in some instances entrepreneurs referred to these instead of to the underlying reasons as the event leading to suspension.

12. By low market potential we mean limited sales possibilities for the program's products under conditions of adequate execution. It is clear that inadequate execution and lack of finance also lead to restricted markets.

either faulty program execution (a result of deficient planning) or inappropriate program selection or uncertainty. Inappropriate program selection might be due to insufficient evaluation of the potential financial (and time) requirements of the program resulting from uncertainties in development and in testing.

5. This refers to (a) unexpected cuts in the financial resources available to the firm or (b) a cut in the budget for the program. Both were present in the seven instances classified in this category of events leading to suspension. In our sample (a) is associated with deficient overall firm strategy [13] or with uncertainty or with both; (b) is associated with deficient portfolio selection.[14]

This category of events can include a world economic crisis, the imposition of tariffs in export markets, the enactment of new safety regulations, and the appearance of competing products. All the events in this category occur after program selection. Deficient behavior in either program selection or execution is explained not in terms of inadequate exploratory search but as inadequate forecasting or monitoring of such events. There is one instance of this category in our sample, the unexpected appearance of a competing product.[15]

### Evaluation of Firm Behavior

Failure is due to either inappropriate firm behavior or uncertainty or both. We focus on inappropriate behavior, at the three levels described in the introductory section.

#### Program execution

There is evidence of inadequate program execution for at least 10 of the 18 failed programs. In 8 (of the 10) there is an almost complete match between a specific event (or events) in category 1 (belated realization of past program requirements) and the type of execution deficiency.[16] This is not a coincidence, since becoming aware of the deficiency (or deficiencies) in execution may in itself lead to suspension. From the point of view of the research, identification of inappropriate program execution necessarily coincides with identification of

13. In the sense of misjudgment of the possibility of raising capital or obtaining loans.

14. Other types of purely financial events may be present in a wider context; e.g., the firm's priorities might shift in response to new threats or opportunities. The links with firm behavior may be more complex than those found here, and a subdivision of the category of purely financial events may be warranted.

15. A total of 30 events leading to the suspension of 18 programs are listed here. This shows that more than one event has led to the suspension of some programs; in such cases it is difficult to trace back inappropriate firm behavior from the events leading to suspension.

16. For some of these eight programs there are also execution deficiencies other than those corresponding to event category 1.

the corresponding event in such cases. The aspects of execution deficiency found in the above-mentioned eight programs can be summarized as follows: *Marketing and promotion effort below required level* (four instances). In these cases, there was no promotion or marketing policy to speak of, the entrepreneur's view being that successful development is tantamount to successful innovation. This made it easier to decide that the effort made was inadequate. *Inadequate spectrum and line of products* (three or four instances). In some cases a very limited set of alternative models of an instrument (spectrum) was offered in situations where user needs are not unambiguously defined and where preferences on working principles vary from user to user (medium or weak market determinateness). In other instances users clearly preferred firms offering a wider range of complementary equipment to firms, such as the ones considered here, offering a narrower range (line).

*Unsatisfactory product specifications (two instances).* We refer here to instances of failure to adapt to user needs beyond what is covered by the preceding paragraph, such as when it is not clear that a need exists at all, when the product is too big and cumbersome and not versatile enough to justify its introduction, or when the need for a very definite specification is not met.

*Inadequate R & D planning* (two instances). In both programs classified here the significance of clinical testing as an integral part of the R & D process—on an equal footing with the electronic or mechanical aspects of development—was not appreciated.

*R & D effort below threshold level* (one instance). Classified here is a failure in a strongly market-determined area where market signals are sufficiently good indicators of development objectives. In these circumstances, lead time (the time from the decision to undertake the program to product launch) can be defined reasonably unambiguously across the set of competing programs and firms. This in turn makes it possible to arrive at a pretty good estimate of the R & D threshold.[17]

An aspect of deficient execution which does not correspond to any of the events leading to suspension is inappropriate selection of agent (four instances). In these cases the firms overlooked the role of the agent in generating acceptance (or firm goodwill) for the product of the programs and in complementing the firm's line. Firms tended to regard the agent as a mere distributor and the agents chosen sometimes had no previous experience of the markets concerned or had no contacts or working experience with potential customers.

Some of the programs associated with deficient program execution suffered from two or three instances of mismanagement. It is noteworthy that most

17. For the present purpose, the R & D threshold is the minimum yearly expenditure in R & D ensuring a sufficiently short lead time to enable the firm to compete in the market.

deficiencies in execution registered are connected with lack of understanding of the market or with marketing (promotion, agent, specifications, and [partly] spectrum) and not with R & D. This reflects the obstacles to success encountered in the Israeli biomedical electronics sector but it may also reflect both the comparative advantage of the investigators in analyzing market aspects and that of the firms in dealing with technology.

### Choice of programs

Inadequate program selection contributed to the failure of at least 11 and possibly 15 programs. It may be useful to distinguish between two types of suspension and failure caused by wrong choice of programs: nonprofitability and the existence of a financial constraint. We define as nonprofitable a program whose private rate of return (adjusted for risk) calculated on the basis of "normal" financial costs is below what can be considered a normal return. A financial constraint exists when the programs are in principle profitable in the sense defined but where the firm does not have the financial means to carry them out successfully.[18]

Firm characteristics to some extent determine whether a particular innovation program failed because of nonprofitability or for financial reasons. Some characteristics, such as the extent of the firm's involvement in the market and its technological expertise, determine the magnitude of the investment required for program execution and thus have a bearing on both the profitability and the financial feasibility of the program. Others, such as the firm's size, influence financial feasibility more than profitability. thus, roughly speaking, the difference between failure due to nonprofitability and failure due to a financial constraint corresponds to the difference between programs which, ceteris paribus, are nonprofitable at all firm sizes and those which are nonprofitable at the firm's present size.

In practice lack of information may make it very difficult to classify failed innovation programs in one of these categories. We therefore decided on a related set of empirical categories for failed programs: (a) programs where there is some evidence of nonprofitability, and (b) programs where the firm does not have the financial means for carrying them through to execution and where there is no evidence of nonprofitability.

In the set of 11 innovations where program selection was inadequate, nine failures were due to financial nonfeasibility (b) and two were nonprofitable. Program selection might have played a part in another four programs; in two

18. The difference between a financial constraint at this level and one at the overall level is that the overall constraint is connected with the firm's set of programs as a whole and not with the specific program analyzed.

of them, failure might have been due to a financial constraint, and in the other two, the program was in theory financially feasible for the firm concerned but success would have required a more concentrated and extensive effort, which did not appear to correspond with firm goals. The evidence on deficient program selection was classified under the following heads: (1) statements by entrepreneurs, (2) qualitative assessments, and (3) other.

*Statements by entrepreneurs.* In 7 out of the 11 programs we have an explicit statement from the entrepreneur that selection was wrong. Such statements are subjective and were made after the program was suspended. Some entrepreneurs find it difficult to admit that a program they had undertaken might have been wrongly selected; insofar as they do admit deficient program selection, they tend to attribute it to a financial constraint and not to inadequate planning. The likelihood of obtaining such a rationalization is high because any innovation involves some measure of market creation, so that failure can always be attributed to inadequate financial backing for this purpose. On the other hand, an ex-post evaluation of past decisions relies on more information than was available at the time of the decision, so there may be a tendency to describe as wrong a selection that was vitiated by the "working out" of uncertainty of deficient program execution even though, given the incomplete information available at the time, it was really correct. Such problems indicate the need for further evidence on deficient program selection.

*Qualitative assessments.* Size and other firm characteristics, on the one hand, and area and program characteristics, on the other, can provide us with some evidence of the financial feasibility of innovations. We would expect the financial requirements of a particular program to be higher (1) the lower the degree of market determinateness of the program, (2) the greater the technological uncertainty surrounding the program, (3) the bigger the system or instrument being developed, (4) the stronger the possible direct effects of using the instrument on the safety of the patient; (5) (associated with [4]) the less known the firm undertaking the program in the corresponding area or market. Factors 1 and 5 are directly related to the magnitude of the marketing effort (including market search and user education) necessary for success (Teubal, Arnon, and Trachtenberg 1976), while factor 2 would be correlated with the required R & D and testing effort. Factors 3 and 4 affect positively both the marketing and the R & D effort required. Also, it must be remembered that the financial capacity of a firm is by assumption directly related to firm size.

These considerations lead to a restricted set of financially feasible innovation programs for new and small firms entering the biomedical electronics sector. The feasibility rule we adopt is as follows: small and new firms cannot financially support innovation programs having two or more of the following characteristics: (1) low or medium market determinateness, (2) high technological uncertainty,[19] (3) big system,[20] (4) a specially sensitive area.[21] This rule indicated that there was a financial constraint in six out of the nine failed programs which we previously classified as being financially nonfeasible. In four (of the six) our conclusion is borne out by the entrepreneur's statement.

The role of market determinateness in determining the financial feasibility and the profitability of biomedical innovation programs should be explored further. Despite more intense competition, choosing programs which are relatively strongly market determinate may be the only possible way of getting started in the business. A significant improvement of function efficiency in standard applications may be rapidly perceived by users without too much effort and the result may be that the firm can soon start selling.[22] An initial success will invite competition from established producers but the firm may be able to reap some profits before being forced to retreat to other areas, and in some cases a policy of forced growth may enable it to withstand competition for some time or even to dominate the market. The point is that a strongly market-determinate program at least has a chance of succeeding, whereas a weakly market-determinate one will probably run out of funds before starting to sell. The former is ceteris paribus, more likely to be feasible than the latter, and it even may be profitable.[23]

*Other criteria.* We include here all other information which may have a bearing on the identification of the set of 11 wrongly chosen programs. Most though not all of this information is numerical: for example, the R & D expenditure estimated by the entrepreneur to be required for completing development or his estimates of threshold R & D and threshold marketing levels. Such estimates were considered in relation to our estimate of the firm's capacity to finance these levels. "Other" information is available on four wrongly chosen programs; in the two cases in which this information suggests financial non-

19. High technological uncertainty implies that the nature of the idea for the new product was such that its theoretical feasibility had to be checked before its technological feasibility.

20. There is a problem in deciding what is a big system. For our purposes a big system would cost the user U.S. $40,000 or more in 1973.

21. Such as some cardiac therapeutic instruments or products whose use involves sending a high voltage through the patient's body.

22. Of course, this will also depend on the firm's reputation, the area considered, the size of the instrument, and user's line requirements.

23. An objection to this statement would be that weak-market-determinate programs have a higher expected rate of return because of lack of competition. While this may be true, the successful introduction of such products will, by reducing the risk for introducing similar products, induce an increase in competition which in turn may reduce the expected rate of return later on. The extent to which this is true depends first on firm characteristics such as size and position in the market and second on area characteristics such as rate of growth of market and extent of potential competition.

Table 2.1. Inappropriate firm behavior and causes of failure

|  | Instances of inappropriate behavior | Inappropriate behavior as cause of failure |
|---|---|---|
| Overall level | 7 | 4 |
| Program selection | 11 | 11 |
| Program execution | 10 | 3 |
| Total | 28 | 18 |

feasibility, it is supported by both the entrepreneur's statement and the qualitative feasibility rule.

### Portfolio selection (overall level)

Inadequate overall firm behavior was traced as contributing to the failure of 7 of the 18 failed programs. Suspension was in all cases due to events in category 5 (financial), reflecting among other things inadequate portfolio selection. Uncertainty undoubtedly also played a part in some cases. The financial resources required for the set of programs chosen by these firms were beyond their capacity. As a result, the programs were suspended before completion of development or at the very early stages of marketing.

### Inappropriate firm behavior and causes of failure

The instances of inappropriate firm behavior discussed above are summarized in table 2.1. As can be seen, many of the failed programs suffered from inappropriate firm behavior at more than one level, and the question arises where the effective causes of failure lie.[24] Consider first situations where there is evidence of inappropriate behavior with regard to program X at both the choice-of-program and the program-execution levels. The behavior at the choice-of-program level should be regarded as the effective cause of failure. Thus of the 10 inappropriately executed programs, 6 were inappropriately selected. In a seventh, suspension was caused by the effects of deficient portfolio selection, which would seem in this case to predominate over deficient execution as a cause of failure.[25] Thus, deficient program execution could be considered responsible for the failure of 3 out of the 18 failed programs studied. Another type of situation occurs when the suspension of wrongly selected programs might have been precipitated by deficient portfolio selection at the firm level. If the choice of program could be unambiguously evaluated independently of

24. In what follows we do not consider the role of uncertainty in causing failure.
25. Execution could have improved if the program had not had to be prematurely suspended.

the choice of portfolio, then in the situation described above wrong choice of program will predominate over wrong choice of portfolio as cause of failure. However, in evaluating program selection, portfolio selection should in general be taken into account, although this is not done by our qualitative assessment rule (see above). In this case, it is not clear which level of mismanagement is the dominant one. Nevertheless the three programs where both levels of mismanagement were involved were considered to have failed because of deficient program selection.[26] This reduces to four the number of programs where the cause of failure lies in deficient portfolio selection. Deficient program selection thus caused the majority of failures, i.e., 11 out of the 18 (see table 2.1).[27]

### Events leading to suspension and causes of failure

We have pointed out some possible links between events leading to suspension and causes of failure. We shall briefly describe the distribution of programs in each event category among the various levels of firm behavior associated with the causes of failure.

Most of the programs (5 out of 8) appearing in the event category "belated realization of past program requirements" and three out of four in the associated categories "belated realization of future program requirements" and "belated realization of potential market" have deficient program selection as a cause of failure. This is to be expected, because if the firm decides to discontinue the program after realizing its requirements, it would not in similar circumstances have undertaken it in the first place had it realized the requirements earlier. A good example is the belated realization that promotion must be undertaken seriously in order to penetrate markets.[28]

The cause of failure in all five programs suspended because of (among other things) "difficulties in the execution of planned tasks" was wrong program

26. In one instance the importance of the program in the portfolio was very great, so that deficient portfolio selection was to a large extent the same as inadequate choice of program. In the two other cases our evaluation is that failure would have occurred even without the negative effect of deficient portfolio selection on program execution. In three other programs affected by deficient portfolio selection not enough information is available for evaluating program selection.

27. The prominence of choice of program may be misleading in view of the difficulties encountered in evaluating inappropriate behavior at this level. Thus the cause of failure may in many cases lie in program execution (or portfolio selection) despite the fact that—to the best of our judgment and on the basis of the limited amount of information at our disposal—the program was wrongly selected.

28. However, when the proximate reason is nonadaptability of the product to user needs, it is conceivable that the development of the original prototype partly depleted the firm's financial resources and thereby reduced its capacity to undertake development in a new direction. In this case, the cause of failure need not be due to wrong program selection but could be due to unsatisfactory execution.

selection. This suggests that the difficulties of execution were largely inherent in the program selected and not so much the result of uncertainty in execution, although this undoubtedly also played a part. Finally the category "purely financial events" is associated with programs where failure was due to inappropriate selection of either program (three cases) or portfolio (four cases). This is certainly what one would expect.

### Interpretation of Firm Behavior

An explanation of firm behavior should begin by indicating the factors influencing decision making and should then proceed to describe the nature of the decision taken by the firm. A central input in decision making is the result of the search activities conducted by the firm. Together with objective characteristics of firms and the orientation of management, these should help to explain actual decision-making patterns.[29] We thus divide the discussion into three parts: a general discussion of search, an analysis of the decision-making patterns of firms, and an attempt to relate these patterns to firm and management characteristics.

#### Exploratory search

Search is defined as information-gathering activity. We are here basically concerned with search which precedes program selection. We call this exploratory search. Broadly speaking there are two types of information that an innovating firm should collect: on the market (including competition) and on technology. Market information is exogenous to the firm and the search for it must usually be conducted outside the firm. Search for information concerning technology, on the other hand, is largely although not exclusively concerned with evaluating the technological capacity of the firm and to that extent is conducted within the innovating organization.[30]

We are here mainly concerned with search connected with the market and not with technology. The results of the search are taken into account when innovation programs are selected and a marketing strategy is formulated for them. It is useful to distinguish between search on user needs and search on the potential market for the products of the program; the first leads to mainly qualitative information and the second to quantitative information.

The object of a search on user needs is to arrive at a dependable translation of those needs into product specifications and design which will maximize the potential market for the product (or more generally, the expected returns from

---

29. Firm characteristics also influence the extent and type of search activity carried out.

30. We are here implicitly assuming a class of science-based firms, which originated with technological know-how or capabilities and not with market know-how or marketing capabilities.

the program). Search on user needs to a very large extent requires personal contact between the innovating firm and potential users, both before and after a prototype has been produced. When market determinateness is high, the appropriate product specification is more easily defined than when it is low. This is because user needs are defined unambiguously to a higher degree of product specificity.[31] When market determinateness is low, user needs are not well defined, nor are they easily perceptible to innovating firms. Search may have to be intensified both because a larger number of potential users must be interviewed and because it is more difficult to interpret their needs in terms of product specifications.[32] Moreover, the proportion of user-need information that is at all searchable is lower (than when market determinateness is high), and correspondingly a greater proportion of the information is obtainable only after market launch.

It seems reasonable to assume that the less market determinate a program, the higher the cost of user-need search, the lower the quality of the information acquired, and the less representative it is. Thus the threshold (minimum) amount of search that must be undertaken in order to attain minimum confidence on the structure of user needs is also higher when market determinateness is low. This may be another reason for small firms not to select weakly market-determinate programs.

Two aspects of market potential must be examined, the potential market for the innovation in general and the share of the market which can be captured by the firm (little will be said about the second at this stage). The former depends both on actual market size (and expected growth) for the class of products associated with the program and on the degree to which the innovation reflects user needs. The latter depends also on firm size and marketing effort and on the intensity, type, and structure of the competition in the markets concerned.

The possibility of estimating the market potential for a new product depends on its market determinateness. In an extreme case where the product is completely new and where consequently no market yet exists, no dependable evaluation can be made. In this case, search on market potential coincides with search on user needs. Quantitative information will have to be extracted exclusively from qualitative observations, and the degree of confidence in such estimates will be low.

The size and expected growth of the market for the product class to which

31. For highly market-determinate products some useful feedback from users is obtainable through professional or trade publications.

32. The information thus acquired is also necessary for a dependable estimation of the promotion and marketing requirements for success.

the innovation belongs represent a starting point for estimating the potential market for highly market-determinate innovations. The market potential could be obtained by estimating the percentage increase in sales derived from improvements in the various dimensions of quality and from price changes. Good estimates of these elasticities are feasible when the structure of user needs is understood, which in turn is feasible when market determinateness is high; moreover, when the innovating firm has been involved in the market for some time, the chances of getting a good estimate improve. Under these circumstances it is possible to choose product specifications reflecting improvements in those dimensions of quality (and price) with a high elasticity, thus maximizing the potential market for the innovation. We thus conclude that the possibility of search on market potential, like search on user needs, is positively related to the degree of market determinateness, as is the quality of the information obtained, while the threshold level of search required is inversely related to market determinateness.[33]

The possibility and costs of obtaining the information required for appropriate decision making also depend on firm characteristics such as firm size and degree of market involvement. The greater the latter, the easier it is to obtain the information and the less it costs.

These considerations constrain the search possibilities of small and new firms, especially in connection with medium and weakly market-determinate programs. This holds both for exploratory search (conducted before program selection) and for search conducted during program execution. It should be noted that all the firms concerned with the 18 failed programs surveyed were new and most of them were small. Nevertheless, 14 out of the 18 failed programs were weakly (or medium) market determinate, and in 13 (out of the 14) there is evidence of insufficient exploratory search.[34] This indicates a certain pattern of decision making based on insufficient exploratory search and on technological management orientation.

### The pattern of decision making

Selection of programs which eventually fail is typically based on an evaluation of the technological feasibility of the programs and an estimate of the R & D costs, with almost no search or evaluation of the potential market for the products of the program and the total costs of launch, especially the costs of

---

33. On the other hand, since higher market determinateness is generally associated with greater intensity of competition, there is a greater need to estimate the share of the total market for the innovation that the innovating firm is likely to capture.

34. In six of these programs there seems to have been no exploratory search at all after the idea of the program had arisen. In general, exploratory search was undertaken in high-and medium-market-determinate programs to a greater extent than in low-market-determinate ones.

promotion and marketing. At the overall firm level the program mix typically chosen by firms with two or more programs is characterized by the absence of marketing synergy between them (some products belonged to widely divergent areas within biomedical electronics) and a very high proportion of low-or medium-market-determinate programs. The financial element of the decision-making pattern is characterized by the inadequacy of the resources allocated to individual programs and by excessive dispersion of resources among competing programs.

What lies behind this type of decision making is the view that the degree of innovation and the marketing (including promotion) effort associated with commercial success are substitutes: the greater the degree of innovation, the lower the market effort required ("the product is so good it sells itself"). This view is unrealistic because of the magnitude of the acceptance problem which characterizes the biomedical electronics sector; this problem and the corresponding marketing and promotional effort required increase with the newness and lack of standardization of the innovation (i.e., the lower the degree of market determinateness). This suggests a distinction between two separate, though interrelated, aspects of innovation, newness and improvement in technical efficiency. The less new a product (i.e., the stronger its market determinateness), the easier it is to measure the efficiency improvement it carries. In the extreme case of the completely new product it is impossible to compare it with others, which may make it difficult to evaluate efficiency improvement in carrying out the associated function.[35] At the other extreme, when the product launched performs existing standard functions more efficiently, the measure of efficiency improvement is fairly straightforward. Our hypothesis is as follows: (1) the newer a product (i.e., the lower its market determinateness), the greater the acceptance problem and the greater the marketing and promotion effort required; (2) the greater the efficiency improvement of a product, the smaller the marketing and promotion effort required.

We can now attempt to explain why firms choose so many medium and weakly market-determinate programs and why all of them were failures. The lack of attention to the market and marketing can be ascribed to their associating or identifying newness with efficiency improvement; this is not, as we have seen, a view that can legitimately be applied to biomedical electronics.[36]

35. This is particularly so for new biomedical instrumentation since its need-satisfaction qualities cannot be easily predicted but can be ascertained only through extensive experimentation.

36. It may be a more adequate description of reality in other areas within electronics where the technical efficiency of new products can be reasonably well predicted in advance, and where experimentation with new instruments does not endanger life. Also, an entrepreneur may subjectively identify the newness of his product with efficiency improvement, but this does not necessarily mean that the market perceives it to be so.

The absence of effective search and the decision-making pattern described here can be explained in terms of the orientation of management, firm size, and extent of market involvement. A small firm is objectively constrained in its search possibilities, while a lack of involvement in the market increases the cost of search and reduces the quality of the information obtained. (For example, market information may be sought from a research man instead of a user.) This will reinforce the tendency of a technologically oriented manager to underestimate the severity of the marketing problem and to undertake programs in which he has a strong professional interest.

## Conclusion

We have presented a tentative framework for analyzing failed innovation programs in the Israeli biomedical electronics sector. The framework is based on a distinction between what we term causes of failure (inappropriate behavior or uncertainty) and their manifestation at the time of program suspension. Inappropriate behavior (and uncertainty) appears at three levels: overall firm level, choice of program, and program execution. Firm variables (size, market involvement, management orientation, and so forth) play a role both in determining inappropriate firm behavior and in explaining it.

Previous related work of which we are aware attempts to cover part of this framework. For example, *On the Shelf* (Centre for the Study of Industrial Innovation 1971) refers to what we term proximate reasons for program suspension (the phrase used there is "causes of program shelving"), and no clear distinction is made between them and the more fundamental causes of failure. On the other hand, phase 2 of project SAPPHO (Rothwell et al. 1974), while attempting to differentiate between inappropriate firm behavior and uncertainty, does not consider the possibility that the former may appear at different levels and deals almost exclusively with what we call inappropriate firm behavior in program execution; this is inadequate when the root of the problem lies in wrong program selection—in fact, the present paper shows that most of the failed programs studied here were associated with elements of inappropriate program selection. The emphasis on program execution reflects the fact that area variables are not included in the main analysis of SAPPHO and that firm size does not differentiate strongly between successful and failed programs. The result has been noninclusion of both sets of variables into the analysis of failure. The main results of our analysis are:

Despite the prevalence of faulty program execution, most failed programs were influenced by inadequate program selection, which should be regarded as a more fundamental cause of failure;

Inadequate program selection is the result of insufficient exploratory search coupled with (and caused by) a management orientation or conception that underestimates the importance of marketing aspects in the innovation process.

Inadequate program execution is often merely a reflection of inadequate selection. Initially it reflects inadequate understanding of the execution requirements for success (that is, a proper understanding would have prevented the firm from selecting the program in the first place); later, once the requirements come to be understood, the program is suspended usually through awareness of the financial constraint facing the firm.[37]

These results may help us decide whether the strong correlation between performance and market determinateness is a reflection of comparative advantage or a result of faulty execution. The fact that in 11 out of 18 programs the cause of failure was faulty program selection (and that in four others, faulty portfolio was also significant) points to a possible pattern of comparative advantage which existed in the Israeli biomedical electronics sector during the period studied.[38] However, the pattern is not so strong as the results of stage 1 would suggest, and this for two reasons: first, in seven programs faulty selection was apparently not the main reason for failure; and the second, the criteria for evaluating program selection are not well defined and the information available is incomplete.

Concerning the latter point, there are two possible extreme cases: (*a*) the criteria for evaluating program selection are clear and there exists full information; (*b*) there are no criteria at all. Suppose also — and this reflects to some extent the industrial experience discussed in this paper — that deficient program selection was found to be the cause of failure in the majority of programs. In case *a* the pattern of correlation between market determinateness and performance would reflect comparative advantage, while in case *b* it would reflect exclusively deficient program execution. The state of the art at present gives us only very imperfect criteria for evaluating program selection. We have something to say about "appropriateness of program selection" while referring fundamentally to program feasibility, but we do not have more refined criteria on optimum program selection. Such criteria as we do have are not unambiguous but probabilistic. Good execution may make a success out of an inappropriate program, although the probability of this occurring is small. In view of this lack of good criteria, we conclude that the prevalence of both inappropriate selection and inappropriate execution implies that while Israeli firms had a comparative advantage in highly market-determinate biomedical electronics programs, this advantage is not as strong as suggested by previous work.

37. The main criterion used here for evaluating program selection was a qualitative rule based on market determinateness and other variables. The former was found in stage 1 to be highly correlated with performance (Teubal, Arnon, and Trachtenberg, 1976), while the latter included variables which were either not highly correlated with performance or not considered (at least explicitly) at that stage.

38. Given the size distribution and characteristics of the firms in the sector.

## References

Teubal, M., N. Arnon, and M. Trachtenberg. 1976. "Performance in Innovation in the Israeli Electronics Industry: A Case Study of Biomedical Electronics Instrumentation." *Research Policy* 5:354–79. (Chap. 1 in this volume.)

Centre for the Study of Industrial Innovation. 1971. *On the Shelf: A Survey of Industrial R & D Projects Abandoned for Non-Technical Reasons.* London.

Rothwell, R., C. Freeman, A. Horsley, V. T. P. Jervis, A. B. Robertson, and J. Townsend. 1974. "SAPPHO Updated—Project SAPPHO Phase II." *Research Policy* 3 (November): 258–91.

This is the second stage of a study on innovation directed by Morris Teubal at the Maurice Falk Institute for Economic Research in Israel in cooperation with the National Council for Research and Development, the Prime Minister's Office, Jerusalem. The first stage is reported in Teubal, Arnon, and Trachtenberg 1976.

We especially appreciate the cooperation of Y. Amir, M. Barone, M. Ben-Porat, S. Ishai, U. Peer, Z. Rosner, Z. Shalev, M. Silberman, A. Suhami, A. Wilenski, and R. Zernik. Although M. Trachtenberg and N. Arnon have not participated in writing this report, their contribution at stage 1 of the study has been of prime importance here. We are grateful to C. Freeman, S. Freund, R. Rosenbloom, R. Rothwell, J. Townsend, and M. Trachtenberg for their comments and suggestions. S. Freund has provided useful editorial help.

# 3 Methodological Aspects of Innovation Research

*Lessons from a Comparison of Project SAPPHO and FIP*

## Roy Rothwell, Morris Teubal, Pablo T. Spiller, and Joe Townsend

Project SAPPHO (Science Policy Research Unit 1972; Rothwell et al. 1974) is an important step in the development of a useful as well as rigorous methodology for innovation research. It systematically attempts to understand and explain commercial success and failure in technological innovation and to consider interindustry differences in the factors involved. A basic characteristic of its methodology is a division of the analysis into two stages: 1, an analysis of factors differentiating successful from failed innovations, and 2, an analysis of factors leading to and explaining failure. These stages are applied to success/failure innovation pairs, 17 of them in chemicals and 12 of them in the scientific instruments sector.

The Falk Innovation Project (FIP) (Teubal, Arnon, and Trachtenberg 1976) was inspired by SAPPHO and resembles it in its basic structure: there is an analysis of factors correlated with commercial performance and this is complemented by an analysis of failure. There are however some important differences between the two studies, which are related to their different objectives: while SAPPHO is basically a study of the *management of innovation*, FIP is an attempt at identifying *comparative advantage* in innovation. This explains the importance in FIP of variables which are largely exogenous to the innovation but which nevertheless determine the requirements for success. These variables, such as rate of growth of market and market determinateness, are termed "area" variables and are not explicitly introduced in the SAPPHO study.

Awareness of these differences led the FIP team to suggest a systematic comparison of both studies. A collaboration with the SAPPHO team was begun at SPRU in the summer of 1975. We here report on the results of the joint effort. The first section presents a fairly thorough comparison of the two studies, and the second section attempts to interpret the SAPPHO interindustry results partly in terms of the FIP area variable of market determinateness. Some methodological implications are spelled out in the third section. They refer to the existence of a hierarchy of variables and the need to postulate subsystems of variables in an analysis of performance. The comparison of the two studies provides a useful background to the proposed approach; in addition, the concept of a hierarchy of variables emerged from the comparison of stage 2. The fourth section reports the results of a pilot test of these concepts on part of the original SAPPHO sample. We conclude by indicating some implications of the analysis for future innovation research.

## Comparison of Methodologies

A comparison of SAPPHO and FIP provides a useful background and introduction to the approach we propose to develop. We start by considering what they have in common: the fact that the universe of innovations studied includes both successful and failed programs (or innovations). In this both differ from earlier studies, such as that of Myers and Marquis (1969), which consider only successful innovations. Thus the main thrust of SAPPHO and FIP is to isolate factors which differentiate successes from failures (or which are correlated with performance), while Myers and Marquis look for factors which are common to a large number of successful innovations. One assumption underlying SAPPHO is that the former set of variables gives a stronger explanation of performance (i.e., success and failure) than the latter, an assumption derived from research in biology. While this assumption should be rigorously examined, there are a priori grounds for regarding factors which discriminate between success and failure as indicating some *effective* obstacle to success; the other set of factors might merely represent characteristics of successful innovations. In short, the interpretation of the significance of each set of variables in explaining performance depends crucially on the underlying conceptual framework, which has not been made sufficiently explicit in either SAPPHO or Myers and Marquis.

Both SAPPHO and FIP felt the need to complement the analysis of stage 1 with an analysis of failure, stage 2. There are two reasons for this: first, the variables which differentiate success from failure across the universe of programs may not be the crucial determinants or "causes" of failure for each program individually; second, it is necessary to explain the causes of failure in terms of other measures (e.g., if failure to understand user needs is a cause of failure, the explanation may be in insufficient search). Thus in stage 1, variables are considered one at a time and are related to the universe of innovations studied, while in stage 2, failures are considered one at a time and are related to the whole or a part of the set of variables or factors. We start with stage 1 and consider two differences between SAPPHO and FIP: first, the type of test carried out to determine whether a particular variable differentiates between success and failure, and second, the type of variables used in the analysis.

In SAPPHO the test with respect to a particular variable is based on a paired comparison of successes and failures. This means two things: first, that each successful member of a pair is compared, with respect to a particular variable, with only the failed member of the pair and not with all other failures; second, that an ordinal relation of the variable is used (i.e., greater or more versus smaller or less). The significance of a particular variable depends negatively on the probability that the pattern of ordinal scores obtained across the innovation set is random.

In FIP the test is based on assessing the level of a particular variable for each R & D program of the set independently. The variable may then be correlated or not with performance. In particular, it differentiates between success and failure if the proportion of one in terms of the other (measured in terms of number of programs or R & D costs) is monotonically related to the level of the variable.

We turn now to the variables used. The paired comparison of SAPPHO is adequate for comparing program execution and firm and management variables because it effectively eliminates the influence of area variables—both programs of a pair belong to the same market and are sufficiently close in time to justify assuming that market and marketing conditions are the same.[1] On the other hand, an across-the-board comparison such as FIP's enables it to consider the role of area variables in performance. As can be seen, each test and the variables analyzed are related to the particular objectives of each piece of research. This also explains why some variables which differentiate between success and failure in one work do not make sense in the context of the other. For example, the variable "amount of marketing expenditures," which discriminates in SAPPHO, would have nothing to say in the context of FIP since the area characteristics of the various programs may imply widely diverging market requirements which may apply to both successes and failures.

In stage 2 (see Teubal and Spiller 1977; Rothwell et al. 1974) the FIP approach used makes two fundamental distinctions: that between the causes of failure (which in general may lie in mismanagement or be due to uncertainty) and their manifestation ("events leading to program suspension") and that between three levels of management: overall firm (the highest level), program selection, and program execution (lowest level). Mismanagement at higher levels often predominates over mismanagement at lower levels. A program-

1. It also ignores the problem of program selection because for every innovation there is a firm which has succeeded and because firm variables do not significantly differentiate between success and failure. This might explain why failure was not related to "wrong" choice of program.

selection feasibility rule has been postulated, embracing the main stage 1 variable (market determinateness) together with other variables (such as size of firm and size of system). The explanation is given in terms of search and the set of considerations entering the firm's decisions. Both are dependent on area variables such as market determinateness (by determining the extent of search requirements) and on firm characteristics (such as market involvement—which determines search possibilities—and management orientation—which determines the extent to which search is considered important).

In SAPPHO the four index variables[2] which significantly differentiated success from failure at stage 1 were automatically considered without further analysis as the causes of failure at stage 2. This has led, as regards mismanagement, to exclusive concern with the program execution level. Consequently, two sets of variables are excluded from the analysis of failure: area variables (since they did not appear in stage 1) and firm characteristics (which in general were not significant in differentiating success from failure). It follows that in a significant number of failures, the cause may be not deficient program execution but deficient program selection—the result of, among other things, area variables being inconsistent with firm characteristics.[3]

SAPPHO's explanation of the causes of failure comprises two steps: first, assigning a set of explanatory measures to each cause, and second, linking each of these explanatory measures to mismanagement, uncertainty, or both.

### Interpretation of Interindustry Differences

The main results of SAPPHO and FIP are not strictly comparable. However, some of the interindustry differences (innovations in instruments versus innovations in chemicals) reported in SAPPHO could be interpreted in terms of the FIP area variable of market determinateness.

In terms of the index variables which best separate success from failure, the most important interindustry differences lie in the relative importance of R & D, marketing, and communications—R & D strength is relatively more important in chemicals (where it ranks first compared with fourth in instruments), while marketing and communications are relatively more important in instruments. Some differences also exist with respect to the index variables of user needs (more important in instruments) and management strength (more important in chemicals).

2. In SAPPHO variables belonging to a particular area of competence (in the terminology used there) were combined linearly to form an index variable. A total of 10 index variables was formed, of which the four that best differentiated success from failure are those corresponding to the areas of competence marketing, R & D strength, user needs, and communications.

3. Neither SAPPHO nor FIP has yet succeeded in devising satisfactory criteria for determining the causes of failure, especially when there is more than one instance of mismanagement at the program execution level.

We first present some specific interindustry differences and then attempt to explain them. A detailed description of the differences between the chemical and instrument innovations reported in SAPPHO can be found in section 5 of Rothwell et al. 1974. We have found it convenient to identify them as follows: (1) search strategy, (2) marketing strategy, (3) R & D strategy, (4) degree of adaptation to user needs, (5) innovation and overall firm strategy, (6) characteristics of the managers of the innovation, (7) firm and organization characteristics.

*1. Search strategy.* Both "market forecasts by sales organization" and "customer involvement during development" differentiate successful instrument innovations from failures. This is not so for chemical innovations.

*2. Marketing strategy.* Successful instrument innovations are usually marketed after the failed member of the pair ("second to market"), while successful chemical innovations are usually first to market. This also reflects some of the aspects of R & D strategy mentioned in the next paragraph.

*3. R & D strategy.* Successful chemical innovations have a shorter lead time, a greater R & D project-team size, and are more radical for world technology than failures, and unlike the failures, they usually undergo periodic program evaluation. Successful instrument innovations have a longer lead time than unsuccessful ones and are less radical for the innovating organizations. The other two measures (size of R & D project team and R & D planning) do not differentiate between success and failure for instruments.

*4. Adaptation to user needs.* The degree to which this occurs is significantly greater in successful instrument innovations than in failures, as can be seen from the variables "modification after prototype" and "bugs in prototype." No such differences are significant for the set of chemical innovations.

*5. Innovation and overall firm strategy.* Firms with failed instrument innovations are less familiar with the markets (i.e., lower synergy) than those with successful ones, while an opposite (although weak) pattern is found within the set of chemical innovations. The share of total R & D spent on the project is also markedly higher in successful instrument innovations than in failures; no such difference is reported for chemicals.

*6. Characteristics of managers.* The successful chemical business innovator has usually been in industry and in the firm longer than the business innovator in the failed cases. He also enjoys higher status and more authority. Such patterns are not characteristic of the instrument innovations—the business innovator of the successful instruments firm has usually spent less time with the firm compared with his counterpart in the unsuccessful firm.

*7. Firm and organization characteristics.* Total employment is higher in successful instruments firms than in failures, and their degree of autonomy from the parent firm (if there is one) is greater. None of these characteristics is relevant for the set of chemical firms. In addition, reorganization of marketing

and nonseparation between development and production differentiate successful and unsuccessful instruments firms; no such thing happens in chemicals.
Finally, internal and external communications, while important in both industries, seem to differentiate success from failure more strongly in instruments
than in chemicals.

In attempting to explain these interindustry differences Rothwell et al. (1974)
concentrated on the characteristics of innovating firms and the characteristics
of users. On the one hand, unlike the instruments firms, the chemical firms are
all large (over 1,000 employed persons), with hierarchical organization, and
this explains the greater role of seniority and power in chemicals and of firm
size in instruments. On the other hand, there are far more users of instruments
than of chemical innovations, a fact which explains why understanding user
needs and external communications are more important in instruments. It is
clear that these differences reflect, among other things, differences in the requirements for innovation success in the two industries.

These requirements are to some extent synthesized in area variables such as
FIP's market determinateness. We also suggest that some other interindustry
differences are related to a variable that can be termed "technological determinateness." Lastly, a theoretical dichotomy between "substances" and "instruments" may help to interpret the area variables.

The concept of market determinateness was developed in connection with
the study of innovations in the Israeli biomedical electronics sector (Teubal,
Arnon, and Trachtenberg 1976, 20–31). The variable indicates the extent to
which needs have been translated into product classes, functions, and specifications. It follows that a better name for the variable is need (rather than
market) determinateness. Thus, the need to reduce costs by 20% is a very
general need with no expression in in terms of products or specifications. But
when the user commissions a particular type of cost-reducing device with
certain characteristics, the need is very specific. In the first case, there is a
high risk of failing to understand the precise structure of user needs, which
in any case may not be very well defined. Moreover, the product acceptance
risk due to the user's incomplete information on the performance characteristics of the product is likely to be very high. These risks diminish when market
determinateness increases. A formulation of this variable in terms of signals
and the translation of the theoretical variable into empirical categories appropriate to biomedical instrumentation were presented in Teubal, Arnon, and
Trachtenberg 1976.

Most of the SAPPHO chemical innovations were process innovations not
involving fundamental changes in existing products, while most instrument
innovations involved a significant new-product component. This suggests that
the chemical innovations were more market determinate than the instrument

innovations, which may in turn explain the greater effort required to understand user needs and the correspondingly greater need for external communications in the instrument industry. It may also explain the smaller role of R & D strength and lead time in performance—in fact, lead times, like R & D thresholds, can be reasonably well defined only when market determinateness is high. Finally, lower market determinateness may also help to explain why instruments firms had to reorganize their marketing in order to succeed and why development should not be separated from production.

The concept of market determinateness suggests a parallel area variable pertaining to technology, technological determinateness, indicating the extent to which existing science and technology determine specific solutions to the technical problems posed by the innovation. Within a particular area, technological determinateness is inversely related to technical risk, while technical risk, together with R & D effort, a firm's familiarity with technology, etc., would determine efficiency in development or lead time.[4] The innovations in chemicals appear to be much more technologically determinate than the instrument innovations: once the specifications of a chemical are established, there are usually only a few alternative reactions that can be developed to achieve them. R & D effort and lead time would therefore be more important factors for performance in chemicals.

Differences in technological and market determinateness between two groups of innovations could be explained in terms of the two being at different stages of the same innovation cycle (e.g., the chemical innovations being at a more mature stage than the instrument innovations). However, each group could be on a different type of innovation cycle. In this connection, a theoretical distinction between two categories of products, "substances" and "instruments," may be useful. One basic distinction between the two would be the number of dimensions expressing user needs and therefore the number of dimensions required to evaluate the degree of adaptation of a product to these needs. This is because a *unit* of an instrument can be defined but a unit of a substance cannot. Thus one characteristic of an instrument would be a large number of dimensions—such as its size and weight, the number of product features, and how they are arranged. Thus, the task of adapting a product to user needs may be much more complex in instruments than in substances, and the extent to which adaptation is achieved may vary more in the former.

To the extent that the chemical innovations in SAPPHO correspond to "substances" and the instruments to the theoretical category "instruments," the explanation given for the interindustry differences observed could be supple-

4. See Teubal, Arnon, and Trachtenberg (1976), for the parallel market-determinateness relations.

mented. However, despite the intuitive appeal of the suggested dichotomy, the concept is at present too hazy to be useful.

### Elements of an Alternative Methodology

A look at the methodologies and results of SAPPHO and FIP suggests that a suitable classification of the variables or measures used is required. There is clearly a fundamental difference in kind between measures such as "after-sales problems" and "diversity of experience of the business innovator" and between these two and market determinateness. This difference should be explicitly defined and introduced in an analysis of performance. Our purpose here is to suggest first a typology of variables. This typology will help us map the requirements for success associated with the environment specific to a class of innovations.

Typology of variables

The variables used in SAPPHO suggest a hierarchy of at least three levels: I, manifestations or results of firm behavior; II, variables related to firm behavior; and III, firm and management characteristics.

Level I variables could also be termed proximate reasons for performance; "understanding user needs" could be an example. Level II variables could, individually or collectively, directly explain the variables of the preceding level. For example, insufficient search on user needs—an aspect of firm behavior— is the immediate cause of a failure to understand user needs. Firm behavior variables also provide the link between firm and management characteristics and the proximate reasons for performance. It is of paramount importance to identify those characteristics of firms and management which are relevant for success in innovation. This requires an explanation of the link between these characteristics and the proximate reasons for performance; and any such explanation should be expressed in terms of firm behavior variables. The distinction between firm behavior and firm management characteristics is also important from the point of view of policy. Firm behavior can, in principle, change in the short run, while firm and management characteristics may be amenable only to gradual changes.[5]

The specific objectives of any particular innovation study will determine whether or not it is desirable to split level II into sublevels. In FIP, a distinction has been made between variables related to program execution (such as marketing effort) and variables associated with program selection. The latter

---

5. At this stage the distinction between variables and characteristics is somewhat fuzzy. For example, "commitment to preconceived design" or "origin of idea" may be regarded either as firm behavior variables or as management characteristics.

include area variables such as market determinateness and strength of competition. In some instances, a study of performance of individual innovations must also refer to aspects of *overall* firm behavior such as the "share of R & D in firm's budget" or the characteristics of the firm's portfolio of innovations.

We now present a preliminary classification of some of the most significant variables of SAPPHO and FIP:

I.  Manifestations or results of firm behavior
    A.  Understanding user needs
    B.  Adaptation by users
    C.  Efficiency in development
    D.  After-sales problems
    E.  Lead time
    F.  First to market

II. Variables related to firm behavior
    A.  Program execution
        1.  Search on user needs
        2.  Customer involvement in development
        3.  Promotion and marketing effort
        4.  User education
        5.  Planning of R & D
        6.  R & D effort
        7.  Spectrum and/or line offered
        8.  Innovation profile
        9.  External coupling in specific technological areas
    B.  Program selection
        1.  R & D synergy with other programs
        2.  Production synergy with other programs
        3.  Marketing synergy with other programs
        4.  Integration with overall firm strategy
        5.  Area variables
            a.  Market determinateness
            b.  Technological determinateness
            c.  Radicalness for world technology
            d.  Strength of competition
            e.  Type of competition
            f.  Rate of growth of market
    C.  Overall firm behavior
        1.  Firm's portfolio of innovations
        2.  Share of R & D in firm's budget

III. Firm and management characteristics
   A. Firm size
   B. Characteristics of firm organization (flexibility)
   C. Internal and external communications
   D. Management orientation
   E. Characteristics of program managers (authority, commitment, and diversity of experience)
   F. R & D strength of firm

### Postulating subsystems of variables

The typology presented above involves a set of vertical links, or interconnections, between variables at different levels. This has implications for the empirical testing of the significance of particular variables, or measures for performance. Some variables (mainly belonging to level I) should be directly correlated with performance, while other variables should be correlated with those at higher levels. This contrasts with stage 1 of both SAPPHO and FIP, in which only direct associations between each variable and performance were tested.

There are two extreme approaches to testing the significance of any one variable or of a set of variables belonging to more than one category. The first approach starts by postulating one or more *subsystems* of potential relevance for performance and then proceeds to test the significance of each. (A subsystem of variables comprises a subset of variables *and* their interconnections. A subsystem should not be postulated unless there is reason to believe that it will be the dominant one for at least one class of innovations.) This would be a structured approach to empirical testing and presupposes some kind of theoretical framework. The other extreme, when no subsystem of variables has been postulated a priori, would be to test all possible associations between variables and between them and performance. This would indicate possible patterns, leading to particular subsystems that permit a more structured approach later on. Classification of variables into levels is a necessary starting point for both approaches. Moreover, it is desirable for the variables within each level to be independent. For the time being we adopt the first approach.

The results of SAPPHO and FIP indicate the potential relevance of two subsystems of linked variables: *A*, which might be called the user-needs subsystem, and *B*, which might be called the lead-time or first-to-market subsystem. The former was seen to be particularly relevant for FIP's and SAPPHO's instrument innovations, while the latter was relevant for SAPPHO's chemical innovations. We postulate a third subsystem, which, like the other two, carries the name of the manifestation variable with which it is linked. It is the efficiency-in-development subsystem (*C*). In terms of the innovation-process

Table 3.1. Three subsystems

| Subsystem *A* | Subsystem *B* | Subsystem *C* |
|---|---|---|
| | *Level I* | |
| Understanding user needs | First to market (or lead time) | Efficiency development (after-sales problems) |
| | *Level II* | |
| Search on user needs | R & D effort | Coupling with outside |
| User education[a] | Investment in first full- | technology |
| Marketing[a] | scale plant | R & D effort |
| Firm's market | R & D synergy[b] | Firm's familiarity with |
| involvement[a,b] | Production synergy[b] | technology[b] |
| | *Level III[c]* | |
| External communications | R & D strength | External communications |
| | | R & D strength |

[a] These variables are especially relevant in a broad view of the user-needs subsystem, where both the determination of user needs *and* the generation of product acceptance are crucial. A good example is biomedical electronics instrumentation (see FIP).

[b] These variables result from the intersection of innovation and firm characteristics and can be traced back to firm behavior (program selection as distinct from program execution).

[c] We have not explicitly written firm size and management characteristics separately for each subsystem, because they are relevant variables for all three. The specific characteristics of management may, however, vary from one subsystem to another; specific tests would have to determine which ones are relevant to each subsystem.

stages proposed by Myers and Marquis (1969)—generation of idea, problem solving, and implementation—the user-needs subsystem would correspond to the market aspects of problem solving, while the lead-time subsystem would correspond to the technology aspects of implementation. The efficiency-in-development subsystem would then refer to the technological aspects of problem solving that are not covered by either of the first two. Thus, the distinction between the problem-solving and implementation aspects of technology and market suggested the efficiency-in-development subsystem. Other subsystems might also be suggested, but seem to be less relevant. The variables composing each subsystem in the context of a SAPPHO-type test[6] are presented in table 3.1.

### The relevance of each subsystem

The relevance of each subsystem in an analysis of performance depends on the general conditions surrounding the universe of innovations being tested. These general conditions are reflected by area variables. At the present time we have

6. I.e., where the object is to test program execution or the management of R & D programs.

only developed one area variable (market determinateness) and we have proposed a second one (technological determinateness).

Our presumption is that the lead-time subsystem is the relevant one if the various competing innovations are reasonably well adapted to user needs. Only then can first-to-market and lower cost be critical factors for success. The conditions for this to occur are strong market and technological determinateness. The former guarantees that the understanding of user needs presents no problems, and the latter ensures that there is no technical problem in translating user needs into the product specifications.

Whenever market or technological determinateness (or both) is weak, there is reason to believe that the competing products will differ in their degree of adaptation to user needs. If market determinateness is weak and technological determinateness strong, the reason will be the unequal perception of user needs by innovating firms, and the user-needs subsystem will be dominant. In the opposite case, the reason will be the high technical risk, which will very likely lead to varying degrees of success in translating clearly perceived user needs into a product. In that case, efficiency in development will be the dominant subsystem.

### A Pilot Test

A pilot paired-comparison test of the subsystems approach was conducted for 12 pairs each of instrument and chemical innovations appearing in SAPPHO, a choice dictated by the availability of the original SAPPHO data and the simplicity of the test compared with one whose aim is to determine comparative advantage in innovation. The specific object was to test the relative strength of each of the three subsystems in each class of innovation.

#### Choice of variables

The set of variables was in the first place determined by the need to test the relevance of each of the three subsystems postulated. This led to a set of 18 variables, to which we added a variable "R & D commitment to the innovation," which could be relevant for any of the subsystems.[7]

#### The matrix of interconnections

R. Curnow of SPRU proposed what turned out to be a very useful tool for determining the relevance of particular subsystems for performance. The matrix

---

7. Except for firm size, where a good direct measure exists, the variables were represented by the SAPPHO measures considered to be the best proxies. In the case of some manifestation (level I) variables such as efficiency in development, a cluster of potential empirical counterparts was examined. Selection of the best proxy was governed by the wording of the question and by its power of discriminating between success and failure.

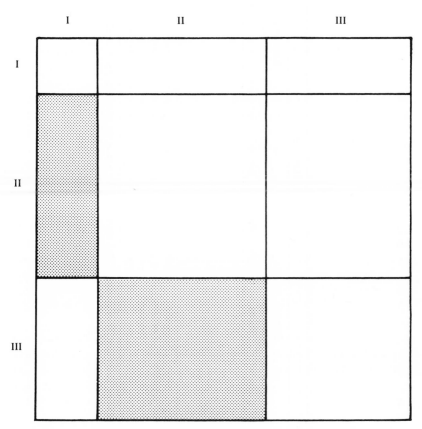

Fig. 3.1. Scheme of matrices of interconnections. Interconnections between variables at adjacent levels were examined (shaded areas), and these parts of the matrix are shown in tables 3.2 and 3.3.

summarizes the associations between the various variables across the innovation set. The theoretical notion of hierarchies, or levels of variables, gave some kind of structure to the kind of interconnections sought. Thus we examined only those between variables at adjacent levels (level II with level I and level III with level II).

Two matrices were built, one for each industry.[8] An example will best illus-

8. Ideally, we would have liked to partition the 24 pairs into four subsets corresponding to the possible combinations of strong (weak) market and technological determinateness, building a matrix for each subset. This was not possible, because of the lack of direct information about the value of the area variables for each innovation program. Little could be inferred about technological determinateness, but there were grounds for believing that market determinateness was on the average stronger for the chemical than the instrument innovations, and this dictated our partition.

trate how the links were correlated. If the successful innovator of a pair both understood user needs better and had more user involvement than the failed member, the pair was said to show a positive correlation between the two variables (recorded as +1). If the ordinal relation was in the opposite direction, −1 was entered. No correlation (0) included ambiguous cases and lack of information. These results were summed, for each pair of variables, over the 12 innovations of the set. The scheme of the matrices is shown in figure 3.1; the relevant parts of each matrix (shaded areas of the scheme) are shown in tables 3.2 and 3.3.

### Empirical test of subsystems

The attempt to derive variable subsystems from the matrix is based on the following principles:

1. A quantitatively significant relationship between two variables was defined as one whose absolute value is greater than or equal to 3. In some instances, however, a quantitative relationship with an absolute value of 2 was included because of its strong a priori plausibility.

2. Any quantitatively significant relationship between two variables that either is based on theory or is plausible (makes sense) intuitively is incorporated into a subsystem. In many relationships found between level II and level I variables the direction can be predicted on theoretical grounds (e.g., a positive relation between R & D effort and development efficiency). This is not necessarily so with relationships between level III and level II variables; for example, it is plausible that there is a relationship between manager's authority or diversity of experience and coupling with outside technology in specific areas, but it is difficult to predict its direction.

3. A subsystem does not show a link between two variables when the relationship is quantitatively not significant or when the relationship is quantitatively significant but makes no sense on theoretical or intuitive grounds. For example, the matrix for the chemical industry (table 3.3) shows quite a strong positive interaction between sales effort and R & D strength. Now these might both be linked via, for example, firm size (more resources available), but any direct causal relationship between the two can be ruled out.

The three subsystems for each industry that emerged from the analysis are user needs, lead time, and after sales. The last could be interpreted as an extended efficiency-in-development subsystem incorporating the variable user education (the link being that user education affects development efficiency as perceived by users, a fact reflected in the variable after-sales problems). It could also be interpreted as a hybrid of the efficiency-in-development subsystem and the user-needs subsystem.

In order to decide in which of the two industries a particular subsystem is more important, the following aspects were taken into account: the degree of completeness of the subsystem, the strength of the significant quantitative links between adjacent levels, and the strength of the link between the level I variable and performance.

A subsystem is considered complete if all the variables to be expected on a priori grounds at each level have significant quantitative relationships with an appropriate variable at the level immediately below. For the most part, level II variables can be assigned to a particular subsystem on a priori grounds. Level III variables are assigned to a specific subsystem via their link with level II variables, but here we often have no a priori reason for expecting particular links.

These criteria are also relevant for determining the dominant subsystem in an industry. However, the criteria for determining the relative completeness of two different subsystems are much more complicated here, because each empirical subsystem must be compared with a different theoretical one. For the time being, only clear cases of dominance can be pointed out.

We now present our results. We start by discussing the dominant subsystem in each industry and then compare the importance of the user-needs and lead-time subsystems across the two industries. The (more problematic) efficiency-in-development subsystem is dealt with at the end.

*Instrument innovations.* In instruments, the user-needs subsystem dominates the lead-time subsystem. A glance at table 3.4 shows that while the lead-time subsystem is exiguous (see also below), the user-needs subsystem is complete (or nearly so). Most of the relevant level II variables exhibit strong links with the level I variable, the only exception being user involvement (+2). Moreover, there is a sufficient number of level III variables with quantitatively significant (and plausible) links to those at level II; they include management characteristics (such as authority), external communications, and firm size. Lastly, the link between understanding of user needs and performance is strong, while lead time practically fails to discriminate success from failure. This empirical result strongly confirms the theoretical prediction.

*Chemical innovations.* As expected, the lead-time subsystem is here the dominant one, since it is more complete than the user-needs subsystem.[9] On the one hand, the lead-time subsystem (table 3.5) exhibits strong quantitative relationships between all relevant level II variables and the level I variable, while the user-needs subsystem (table 3.6) has only one relationship of this kind. On the other hand, hardly any plausible relationships appear between level III and II variables in the user-needs subsystem. This is clearly not so for lead time.

9. Despite the fact that the relationship between first-to-market and performance is *slightly* weaker than that between understanding of user needs and performance.

Table 3.4. Instruments industry: two subsystems

| Performance | Level I | Level II | Level III |
|---|---|---|---|

*Lead time*

Succeeded

4 firms ⎱ First
3 firms ⎰ to market     (+3) R & D commitment

Failed

*User needs*

(+5) User education ◄────── (+2) Firm size[a]

Succeeded

9 firms ⎱ Better
0 firms ⎰ understanding of user needs ◄── (+5) Sales effort

(+3) Manager's authority

(+3) Manager's diversity of experience

Failed

(+3)
(+4) Firm size[a]

(+2) User involvement     (+3) External communications

(+4) Manager's commitment

[a]Firm size is linked with all three level II variables shown and appears twice under level III arrows. The same system is used in subsequent tables.

*The user needs subsystem.* In chemicals, only user education (+3) is quantitatively significant at level II (table 3.6) while in instruments this variable and sales effort both show +5 (table 3.4). Similarly, the instruments industry exhibits several plausible and significant relationships between levels III and II (e.g., firm size and manager's authority are significantly related to sales effort). This is not so in chemicals.[10] We conclude that (as expected) the user needs subsystem is more complete, and hence more important, in instruments than in chemicals.

*The lead-time subsystem.* The lead-time subsystem (table 3.5) is more important in chemical than in instrument innovations. In chemicals, it is nearly

10. In chemicals, there is a relationship of +3 between manager's authority and user education, but the reason for it is not clear.

Table 3.5. Lead-time subsystem: both industries

| Performance | Level I | Level II | Level III |
|---|---|---|---|

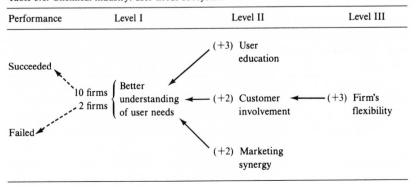

Table 3.6. Chemical industry: user-needs subsystem

| Performance | Level I | Level II | Level III |
|---|---|---|---|

Table 3.7. Instruments industry: efficiency in development (after-sales problems)

| Performance | Level I | Level II | Level III |
|---|---|---|---|

complete, which is not true for instruments. In the latter, there are no significant quantitative relationships between lead time and the R & D effort and the synergy variables, while these relationships do appear in chemicals. The only significant level II variable in instruments is R & D commitment (which also appears in chemicals), but it is not significantly linked to any of the management variables of level III. By contrast, the lead-time subsystem in chemicals exhibits a strong relationship between management authority and both R & D commitment and R & D effort (+5 in both cases). Last but not least, the relationship between the first-to-market variable (shorter lead time) and success is much stronger in chemicals than in instruments (where in fact, the first-to-market variable does not appear to discriminate at all between success and failure).

*The after-sales subsystem.* Since efficiency in development was not measured

Table 3.8. Chemical industry: efficiency in development (after-sales problems)

| Performance | Level I | Level II | Level III |
|---|---|---|---|

Succeeded

Failed

7 firms
0 firms

Fewer after-sales problems

(+3) User education ← (+3) Manager's authority
(+2) authority

(+2) Manager's commitment

(+3) Coupling in specific areas

(+2) Manager's diversity of experience

(+2) External communications

(−2) / (−3) Firm size

(+3) Production synergy ← (+3) Manager's authority

(−3) R & D strength

(+2) Customer involvement ← (+3) Firm's flexibility

directly, we had to select a proxy for it from the SAPPHO variables. After-sales problems was considered to be the best, despite the fact that it is not clear whether the variable reflects development inefficiencies or misunderstanding of user needs. This problem can be tackled by looking at the level II variables related to after-sales problems. If most of the quantitatively significant variables are nonmarketing ones, such as R & D effort, R & D and production synergy, and coupling with outside technology, we have an indication that after-sales problems is a good proxy for efficiency in development. If, on the other hand, the quantitatively significant variables are mostly marketing variables (such as user education, customer involvement marketing effort and synergy), the implication is that after-sales problems result mainly from lack of understanding of user needs. In this case, there is no point in considering a separate after-sales subsystem.

As can be seen from table 3.7, this is the situation in instruments. In chemicals (table 3.8) it makes sense to consider the after-sales subsystem separately since nonmarketing variables appear quite strongly at level II. The after-sales subsystem seems to be more complete than the user-needs subsystem. It is also less important than the lead-time subsystem, both because it is less complete, particularly owing to the absence of R & D effort at level II, and because the links between adjacent levels are weaker than in the lead-time subsystem.

### Comments and Conclusions

SAPPHO's and FIP's methodologies were compared and the SAPPHO results reinterpreted in terms of FIP's main area variable, market (or need) determinateness.

The comparison suggested a new approach for studies of innovation performance. The approach has two elements: a hierarchy of variables composed of three levels and subsystems of interrelated variables. We postulated three subsystems, user needs, lead time, and development efficiency. Any one of the three could, on a priori grounds, have been dominant, depending on the circumstances reflected by the level of the area variables.

The pilot test on 24 SAPPHO pairs (12 instrument and 12 chemical) gave encouraging confirmation to the predictions, but it is also clear that much work has yet to be done in developing the methodology so as to arrive at a useful analytical and operative tool. There are four areas where a special effort is needed.

*Development of key area variables.* The variable market (or need) determinateness should be adapted to the circumstances of other classes of innovation, such as those of the drug and mechanical engineering industries. A still greater effort should be made to arrive at a satisfactory conceptualization of the variable of technological determinateness and its relation to technological risk. Lastly, despite its undoubted importance, nothing was said about competition, because as yet no meaningful conceptualization of strength and type of competition has been achieved.

*Development of subsystems.* Further work should be devoted to refining the various subsystems proposed, both as regards their structure and as regards conditions for dominance.

*Testing the approach.* The approach requires further testing on existing bodies of data. Existing statistical methods should be screened and possibly adapted for this purpose.

*Comparative advantage in innovation.* The methodology should eventually be extended to make possible studies of comparative advantage in innovation. Such studies would involve at least two stages: first, testing the appropriateness of innovation programs by using a series of area variables in the light of characteristics of the firm and the environment (test of program selection); second, evaluating the execution of programs (test of program execution). At present the analytical framework is inadequate for this purpose.

# Appendix

Proximate Variables Selected from the SAPPHO Coding Sheet
and the Levels to Which They Were Assigned

| SAPPHO/FIP variable for pilot test | SAPPHO question and question number on questionnaire |
| --- | --- |
| Level I. Manifestations or results of firm behavior | |
| Understanding user needs | Were user needs more fully understood by the innovators in one case than in the other? (143) |
| After-sales problems | Were there any after-sales problems? (145) |
| Lead time (first to market) | What was the date of the first commercial launch? (1) |
| Level II. Variables related to firm behavior | |
| A. Program execution | |
| Customer involvement | Were potential users involved in development at any stage? (138) |
| User education | Were any steps taken to educate users? (141) |
| Sales effort | Was the sales effort a major factor in the success or failure of the innovation? (147) |
| Coupling in specific areas | What was the degree of coupling with the outside scientific and technological community in the specialized field involved? (78b) |
| R & D effort | How large a team was put to work on the innovation at the peak of the project? (60) |
| Commitment to program | What percentage of existing R & D manpower was allocated to the project at commencement? (16) |
| B. Program selection | |
| R & D synergy | Was the innovation more or less radical for the firms concerned? (2a) |
| Marketing synergy | Did the innovation involve going into an unrelated area from the point of view of the market? (132) |

## Appendix (continued)

| SAPPHO/FIP variable for pilot test | SAPPHO question and question number on questionnaire |
|---|---|
| Production synergy[a] | If it was a product innovation, was the main manufacturing process needed established in use or familiar to the firm more in one case than in the other? (118) |
| | If it was a process innovation, did it involve radically new equipment unfamiliar to the firm more in one case than in the other? (119) |

Level III. Firm and management characteristics

| | |
|---|---|
| External communications | Did one organization have a more satisfactory communication network than the other, externally? (36b) |
| R & D strength | How many employees did the (main formal) department have at the beginning of the project? (44) |
| Firm size | In terms of the organization coded under question 28a, what was the total employment at the commencement of the development?[b] (29) |
| Flexibility of the firm | Was the organization deliberately modified at any time to improve the possibilities of this innovation? (37) |
| Project manager's authority | Did the business executive[c] have more or less authority (power) in one case than in the other? (96) |
| Project manager's commitment to program | Did the business executive[c] have more commitment (enthusiasm) in one case than in the other? (97) |
| Project manager's diversity | Did the business executive[c] have a more diverse experience in one case than in the other? (89) |

[a]Question 118 was used in the case of a product innovation and question 119 in the case of a process innovation.

[b]Question 28a: What kind of organization marketed the innovation? 1 = independent industrial firm; 2 = subsidiary firm; 3 = individuals; 4 = independent new organization; 5 = controlled new organization.

[c]The business executive was the person actually responsible within the management structure for the overall progress of the project.

### References

Myers, Sumner, and Donald G. Marquis. 1969. *Successful Industrial Innovations.* National Science Foundation NSF 69-17. Washington, D.C.

Rothwell, R., C. Freeman, A. Horsley, V. T. P. Jervis, A. B. Robertson, and J. Townsend. 1974 "SAPPHO Updated — Project SAPPHO Phase II." *Research Policy* 3 (November): 258-91.

Science Policy Research Unit, University of Sussex. 1972. *Success and Failure in Industrial Innovation: Report on Project SAPPHO.* London: Centre for the Study of Industrial Innovation.

Teubal, Morris, Naftali Arnon and Manuel Trachtenberg. 1976. "Performance in Innovation in the Israeli Electronics Industry: A Case Study of Biomedical Electronics Instrumentation." *Research Policy* 5:354-79. (Chap. 1 of this volume.)

Teubal, Morris, and Pablo T. Spiller. 1977. "Analysis of R & D Failure." *Research Policy* 6:254-75. (Chap. 2 of this volume.)

We appreciate the comments of R. Curnow, C. Freeman, S. Freund, and M. Trachtenberg.

# 4 On User Needs and Need Determination
## Aspects of the Theory of Technological Innovation

## Morris Teubal

Technological innovation, like the activity of production, may be regarded as induced by human needs, but unlike production these needs are frequently not represented by an unambiguous and well-defined market or demand curve. Innovations generally involve a new-product component, and insofar as this is so, they *precede* the generation of markets and demand curves. They should accordingly be regarded as responses to more general, less-defined needs than those expressible in terms of well-defined markets or demand.

Economic theory has yet to develop a theory of needs which can effectively explain both the commercial performance of innovations and the rate and direction of inventive or innovative activity.[1] Despite this, concepts such as user needs and need determinateness have been introduced in studies on innovation performance. For example, project SAPPHO, which explains commercial success and failure in a set of chemical and instrument innovation pairs, has concluded that successful innovators systematically understand the needs of users better than failed innovators (see Science Policy Research Unit 1972). Another study, on the Israeli biomedical electronics sector, has concluded that all the commercially successful innovations of the sector were highly need-determinate, whereas most failures were not (Teubal, Arnon, and Trachtenberg 1976).

The use of the notions user needs and need determinateness has, as mentioned, preceded their rigorous conceptual clarification. Although they were found useful and even necessary in empirical studies, further progress is contingent upon such a clarification. This is attempted in the following two sections of this paper. The third section deals with the process of need determination and leads us to a discussion of product (or innovation) cycle theory. The relationship of the need determination process to the processes of product

1. The characteristics approach to consumer behavior is an important first step in this direction. See Lancaster 1971; and Ironmonger 1972.

standardization is discussed, and some implications for product competition, comparative advantage, and business strategy are indicated.

### User Needs

The terms *user needs* and *product adaptation to user needs* refer to product quality. The context in which they are used in innovation studies involves first and foremost what may be termed the choice of quality, which means the choice of one product option from a set of options.[2]

Consider some broadly conceived need such as data processing that is to be satisfied by a service whose extent depends on specific properties such as accuracy, reliability, efficiency, versatility, safety, and simplicity of operation. These properties correspond to (abstract) product performance dimensions; e.g., accuracy is both a property of the service and a performance dimension of products. A product's quality is the extent of service provided by it: it therefore depends on product performance dimensions in the same way as the extent of the service depends on the properties of the service.

It is assumed that performance dimensions cannot in general be added across units of a given product (even when an ordinal ranking is possible). User needs in general are the preferences of the user for the properties of the service or alternatively for the performance dimensions of the product. It is assumed that these have minimum levels for the service provided by a product (or for product quality) to be positive or for the product to be considered at all for purchase.[3] This aspect of preferences will be termed *minimum user needs*. Finally, while the term *product adaptation to user needs* used in innovation studies is synonymous here with product quality, a narrow definition could also be postulated. Under this definition a product would be said to be adapted to user needs if two conditions hold: (1) it satisfies minimum user needs; (2) its price is below the maximum price the user has assigned to satisfying the need under consideration.[4]

#### Purchasing decision

Under perfect certainty the user's decision depends on the quality and the price of each element in the set of product options adapted to his (minimum) needs. His first task is therefore to determine this set of product options.

An example will clarify this point. Say a doctor wants to purchase X-ray

2. For a related viewpoint see Sweeney 1974, 147–67.

3. E.g., there are minimum safety standards for an airplane to be considered fit for carrying passengers. Sometimes these minimum levels are considered part of the definition of the product class.

4. This maximum price cannot exceed the user's resources. Thus a given product may be adapted to users belonging to one market segment (e.g., "rich" users) and not to the users belonging to another (e.g., "poor" users).

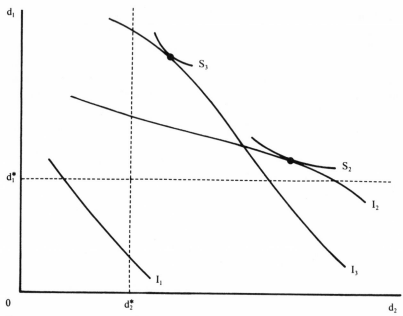

Fig. 4.1. Service (or quality) ranking of products.

equipment with the relevant minimum performance dimensions corresponding to patient safety $d_1$ and accuracy in measurement $d_2$ ($d_1^*$ and $d_2^*$, respectively). Beyond these minimum levels the doctor has some preference which may be expressed in terms of a service function $S(d_1, d_2)$, where $d_1 > d_1^*$, $d_2 > d_2^*$. These preferences are drawn in figure 4.1 (schedules $S_2$, $S_3$).

Suppose that three instruments, $I_1$, $I_2$, and $I_3$, are offered in the market and that the price of each is below the maximum the doctor is willing or able to spend. Each instrument is characterized by a particular trade-off function between $d_1$ and $d_2$.[5] From the figure it is clear that the set ($I_2$, $I_3$) is the collection of instruments satisfying user needs. Instrument $I_1$ will not even be considered for purchase. The issue is now to decide between purchasing $I_2$ or $I_3$. The decision will depend not only on the service obtained in each case, $S_2$ and $S_3$, but also on the respective prices, $P_2$ and $P_3$. If $P_3 > P_2$, then $I_3$ should be balanced against the value of the additional expense that the doctor must incur

5. A special case of the function would be a single point. In some instances a point in ($d_1$, $d_2$) would be a better representation of a product option than a schedule; in others, a schedule might be more realistic, e.g., trade-offs between speed and safety in cars.

The extent to which the performance dimensions of instruments can be specified depends on the degree of need determinateness (see below). When this is low the I schedules cannot be traced with certainty as has been done in the figure.

to purchase it. This necessitates recourse to a higher-level utility function, one argument of which will be *S*.

In short, the purchasing decision process here described differs from the usual ones is some interesting ways:[6] (1) The *basic* decision is to purchase *one* of a set of alternative instruments to the exclusion of all others, and not varying quantities of one or more alternatives.[7] (2) The decision-making process comprises at least two stages: first, when a set of options adapted to (minimum) user needs is determined, and second, when the choice within this set is made. (3) The decision at the second stage is not based on marginal analysis but on a discrete comparison of the service provided in relation to the cost of the alternative options within the set of products satisfying (minimum) user needs. (4) A hierarchy of utility functions may have to be introduced in order to arrive at a purchasing decision.

### Need Determinateness

Need determinateness is the extent to which preferences are specified (or need satisfaction is expressed) in terms of product classes, functions, and features. At one extreme there are needs which are very general in the sense that they do not point to any particular product class as providing the corresponding service. Examples of such general needs are national defense or even data processing without further specifications. At the other extreme we find needs which have been specified in great detail, such as those leading to the commissioning of specific items of new military hardware.

It may be useful to distinguish three degrees of need determinateness:

*Identification of product class.* This refers to the (eventual) association of the basic need with a particular product class (or classes), for example, computers in relation to data processing after World War II.

*Specification of user needs.* This is the determination of the relevant abstract performance dimensions (such as speed and accuracy) and the way they affect the service provided.

*Translation of user needs.* By this we mean stating user needs in terms of concrete product characteristics. This implies relating concrete product characteristics to abstract product performance dimensions.

We suggest three degrees of specificity: product functions (e.g., multiplication, division, etc., in a calculator), product features (e.g., the presence or absence of power steering in automobiles), and product specifications (such as the array of parts or components and their dimensions). This hierarchy of

---

6. These were developed jointly with M. Trachtenberg.

7. This would not apply to homogeneous commodities, which we call *substances*. Most of this paper, however, deals with products such as instruments or machines, which we call *systems*. For a discussion of this distinction, see below.

concrete product characteristics, or product aspects, is to some extent arbitrary and its relevance should be determined for each product class (see fig. 4.2). Increasing need determinateness would here involve a relationship between performance dimensions and increasingly specific product aspects. This relationship involves both qualitative and quantitative elements. The former include the product aspects at a particular level required to infer more general product aspects (e.g., a check-digit verifier, which ensures the accuracy of data entry into an electronic data-processing system). Quantitative elements enter into efficiency considerations: the efficiency of functions will depend on the characteristics of features, some of which are quantitative (e.g., size of a TV screen for efficiency in data display); the characteristics may also have quantitative threshold and saturation levels.[8]

### Need determinateness and innovation performance

The innovation process is characterized on the market side by at least two types of uncertainty: the uncertainty of innovator firms about user needs and their concrete translation; and the uncertainty of users about product quality. These uncertainties confront innovator firms with two problems: understanding (and specifying) user needs, and gaining product acceptance. Success with the latter involves convincing users to include the product offered by the firm in the set of alternative product options.[9] Under perfect certainty of quality evaluation and leaving aside problems of product development, we would only have the first problem: once this is overcome, products would be adapted to user needs and automatically accepted. Acceptance appears as a separate problem once uncertainty of quality evaluation is postulated: the position of the $I$ locus of a particular good is not known to the user with certainty. As in the perfect certainty case (fig. 4.1), he will determine a set of product options from which the choice is to be made (the set of accepted products), but this set will not necessarily coincide (as it does in the certainty case) with the set of products (objectively) adapted to his needs.[10]

8. We should distinguish the degree of need determinateness from the process of need determination. The process in general is a complex one. Some aspects of the translation of user needs normally emerge before their full specification. Moreover, some performance dimensions (and even product characteristics) are implicit in the product class identified: e.g., the speed of automobiles.

9. Product acceptance is thus a *necessary*, but not a sufficient, condition for demand; acceptance of a product may in fact precede the purchasing decision, and at the time of that decision any accepted product will have to compete with other accepted products of the set of purchase options. See Teubal, Arnon, and Trachtenberg 1976.

10. The user cannot ascertain product quality but will attempt to evaluate or assess it. His assessment will be both direct and indirect (via proxies for quality). The user's need determinate-

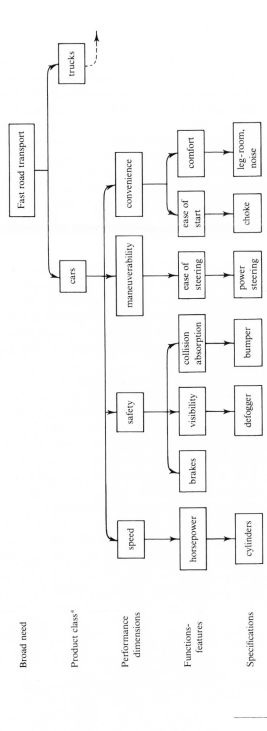

Fig. 4.2. Hierarchy of product aspects.
[a]Overall characteristics such as range of sizes and of carrying capacity are determined by the product class identified.

An increase in need determinateness reduces both types of uncertainty: the more specific needs will be signaled to firms, and they also represent more specific standards for evaluating quality. Correspondingly, the magnitude of both problems confronting innovator firms will decline. We thus have:

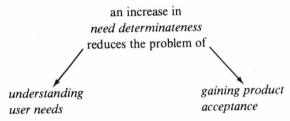

an increase in
*need determinateness*
reduces the problem of

*understanding*
*user needs*

*gaining product*
*acceptance*

This has the following implications for innovation performance: the two conditions required for commercial success are (1) the launching of a product which is objectively adapted to user needs and perceived as such by potential users, and (2) the launching of a product which can stand up to competition from other firms.

As we just saw, an increase in need determinateness reduces the effort required to fulfill the first condition, thereby bringing the second condition to the fore.[11] This means a reduction in the importance of those informational aspects of marketing with a feedback element, such as search on user needs, product testing, product demonstration, and user education. On the other hand, it increases the importance of the R & D effort required to make the product competitive and the marketing effort associated with distribution.

### Relation to consumer demand theory

The analysis described up to now will be related to developments in demand theory that make possible the consideration of demand for new goods, namely, what is known as the characteristics approach to consumer behavior (Lancaster 1971).

The aim of Lancaster's (1971) work was to construct a theory of consumer behavior based on product characteristics rather than goods, and the signifi-

---

ness and his past history of using similar products will affect his direct assessment of product quality (see the Appendix). Possible proxies include things such as user experience with the firm, firm reputation, and the opinion of other users.

11. Alternatively, we may say that when need determinateness is high, the major problem is to compete successfully with other accepted products, whereas designing a product which users will include in their set of alternative product options will present less difficulty. When need determinateness is low, the latter is the major problem facing the firm; thus the absence of suitable proxies for quality may explain the difficulties encountered by small and new innovating firms when launching new products. See Teubal, Arnon, and Trachtenberg 1976; Jones 1969.

cance of his theory for innovation performance is that it provides a framework for estimating the demand for a new good. An assumption which is crucial in our context is that the product characteristics directly affecting utility are concrete and observable. Two further assumptions follow (though not only from this): that consumers are able to define their preferences and that they are able to evaluate product quality with certainty. For if utility depended on abstract unobservable product-performance dimensions rather than on concrete observable product characteristics, consumers—even if able to define utility in terms of the abstract dimensions—would not automatically be able to define their preferences completely in terms of concrete product characteristics. A relation between the concrete characteristics and the abstract dimensions would have to be postulated (for the same reason quality evaluation would not automatically result from checking or listing the set of concrete product characteristics).

The procedure for estimating the demand for a new product, given its price and characteristics, assumes that the set of related products and their prices remain unchanged after the new product is launched. In order to explain the commercial performance of innovations, the procedure must take into account at least two further considerations: (1) The new product associated with an innovation is only one of a range of possibilities, each with a different set of characteristics (objective and perceived). (2) The estimated sales of a new product will depend on the innovation strategies of competing firms, including the competitive reactions to the product considered, and on the growth of the total market for related goods.

The range of alternative products that may be launched by a particular innovation program will depend on the product class involved and on the R & D expenditure contemplated. Given these and ignoring technological risks, the probability that the objective characteristics of the choice will best reflect user needs and moreover that they will be perceived as doing so by users depends on the extent of need determinateness and on the firm's marketing effort. Innovation performance will then depend, in the first instance, on these factors. But there are other factors as well. The first is actual competition, by which we mean the "distance" between the price and characteristics of existing products and those of the new product. Given the distribution of preferences and incomes in the relevant range, this is all that is required when market size is constant and when no other firm innovates.[12] Demand theory is certainly very useful at this point. The second factor is potential competition (in relation to market growth). Competing firms will normally be launching new products

12. As when estimating the demand for new transport modes commissioned by government agencies.

or improving existing ones, sometimes in response to the new product. We have as yet no theory of potential competition which could be used in conjunction with a theory of needs and a theory of demand in order to explain and predict innovation performance. There is no doubt that refinement of the notion of technical uncertainty and further developments in the theory of firm strategy are required for this purpose.

### The Process of Need Determination

We have yet to describe the process of need determination. At least three aspects are involved: user experience, user adaptation, and user reaction to new product-price options. We shall deal briefly with each of them.

In the context of a given product class, need determination resulting from pure experience may be visualized as a sifting out from a given set of options —by using the products—a set of product features which are "relevant" for need satisfaction. Need determination will in general be associated with the generation of a clearer ranking of preferences with respect to the set of product options considered. But this is not sufficient: need determinateness might have failed to increase because the user has not linked need satisfaction to a definite set of product features. The reasons for this may include a lack of knowledge on the part of the user and insufficiently strong differences in preferences for the various options so that even product features associated with the less-preferred options may be considered relevant to need satisfaction.

When the magnitude of user adaptation to a product (i.e., complementary investments by the user required to ensure adequate service) is significant and specific to a particular model or firm, then products currently in use will define the preferences for additional or subsequent units. This might entail need determination, e.g., when no clear preferences existed before the original purchase and when the preferences emerging later are associated with definite product features. An example in the electronic data-processing market might be the preference of IBM customers for computers capable of using IBM software. Another example comes from the printing industry: "Any attempts to change the layout of keys on a typewriter have always met with stubborn resistance because of the skilled pool of labour already familiar with the 'universal board' and because the changeover would be extremely expensive and troublesome."[13]

A new product-price option appearing in a situation of user indifference between existing options of a set may trigger off need determination. What is required is the emergence of a strong preference for the new option, that the

13. "Filmsetting in Focus," *Monotype Recorder* 43 (Summer 1965): 25. My attention was drawn to this example by S. Freund.

preference be associated with distinct product features, and that the features should be linked by users to the significantly higher service generated.[14] An example from nuclear medicine is the improvement in the resolution of the gamma camera, as a result of which there has been a shift from competition between it and the nuclear scanner to a clear preference for the camera in the diagnosis of most organs.

The availability of new options is the result of technological innovation, both product innovation leading to new product features and process innovation leading to reductions in product price. Thus technological innovation is central to need determination, although it certainly acts in conjunction with user experience and user adaptation.

### The product cycle and product standardization

We have seen that need determination results from the interaction between needs and products (or more fundamentally, technology) and is thus one of the several processes unfolding along the product, or innovation, cycle. This interaction is the unique feature of the product cycle; otherwise, standard supply and demand analysis would suffice, the cycle being a mere sequence of pairs of supply and demand curves through time with each pair independent of the others. The interaction between supply and demand can be interpreted in terms of a learning process: consumers learn what they want or need; producers learn how to innovate. This process involves not only quantitative elements but qualitative ones as well: the products or characteristics desired, the technologies and process equipment required, etc.

A central proposition in the standard product-cycle description is the tendency for increasing product and process standardization (Vernon 1966; Hirsch 1967). This tendency is then related to changes in the patterns of competition (stronger, more based on price), in the comparative advantage of firms and countries (greater importance of price owing to economies of scale in production), and in business strategy (a progressive reduction in the rate of innovation) (Utterback and Abernathy 1975). Walker (1975) has challenged this assumption and pointed out cases of decreasing product standardization, e.g., in textile machinery. This alternative pattern seems to be more and more relevant in industrial societies where a new technology developed for a particular product eventually invades other areas—the classical example being the invasion of the textile sector by the chemical industry. Now if this is true,

---

14. A special case of reaction to a new option would be exemplified by the appearance of a low-cost version of an existing product which draws a (new) set of low-income users to consider the product for purchase. They would have been indifferent to the original options in the sense that they could not have considered them for purchase because of the price.

then there are far-reaching consequences for competition, comparative advantage, and business strategy.

What is then required is a *theory* of the product (or innovation) cycle which should, first, explicitly analyze the conditions for increasing standardization[15] and, second, postulate a classification of innovation cycles that would be relevant when analyzing competition, comparative advantage, and business strategy. We first discuss product standardization and link it with need determination. We shall see how this requires explicit treatment of technological change and how it is closely linked to the emergence of submarkets, or market segments. We end up indicating two different kinds of product cycle. Throughout, implications for product competition, comparative advantage (innovation performance), and firm strategy are pointed out.

### Increasing standardization

We define increasing standardization as a process by which increasingly specific product aspects become similar across firms. For example, the only element initially common to the various products offered to satisfy a need is their belonging to some product class (low standardization); then the basic functions become similar, and eventually product specifications become identical (maximum standardization). In the first situation, *all* product aspects (i.e., performance dimensions, product functions, and product features) are likely to differ among products (firms); in the last situation, no product aspect differs.

It is useful at the outset to present a set of sufficient conditions for increasing product standardization. These conditions are rather artificial and will be relaxed later on. No explicit treatment of need determination is given at this stage. The set of conditions is the following: (1) increase in need determinateness, (2) identical users,[16] (3) static technology, (4) perfectly malleable technology, and (5) user-oriented firm strategy.

An increase in need determinateness is a necessary, but not a sufficient, condition for product standardization to occur. It does mean that firms receive increasingly specific need signals from users but does not automatically imply increasing standardization of products; either some (or all) firms do not want to respond to the signals, feeling that they can introduce more revolutionary products whose importance users do not presently recognize,[17] or firms differ considerably in their success in developing products that incorporate the equally perceived, increasingly specific need signals.

15. This has not yet been done systematically in the literature. Explicit consideration of need determination may be an essential requirement.

16. This assumption for the time being excludes the emergence of submarkets (e.g., the commercial and scientific submarkets within the computer industry) from the discussion.

17. This would be a technology-oriented firm strategy.

By a user-oriented strategy we mean that firms will attempt innovations (*a*) that incorporate those product aspects for which clear user signals are forthcoming, according to the degree of need determinateness, and (*b*) that have a random pattern of characteristics for more specific product aspects. By technology we mean in this context the set of pieces of knowledge or information concerning product development relevant to the cycle under consideration (we are here not concerned with process technology). A static technology means that this set is constant; i.e., there are no additions or subtractions of relevant knowledge. This ensures that there is an optimum way of attaining particular technical objectives which remains invariant. By perfectly malleable technology we mean that the pieces of knowledge forming the technology can be drawn upon by every firm and combined with other pieces of knowledge without cost (and instantaneously). Since an innovation in this context is a new combination of pieces of knowledge, this means that innovation costs are zero for all firms and that any firm will be able to achieve its technical objectives as successfully as any other firm. The assumptions of static and perfectly malleable technology rule out need determination induced by reaction to new options and to user adaptation, leaving only user experience.

The result of these assumptions is that all firms will put out products which are the same with respect to aspects for which clear signals exist and which are likely to differ with respect to more specific aspects. In this way need determination will necessarily induce product standardization.

### Emergence of submarkets

When the "identical users" assumption is relaxed, we can incorporate the emergence of submarkets in the model, each consisting of a set of users defining and determining their needs in the same way, which differs from that of other users. The concept of submarkets is especially relevant when we also relax the assumptions about technology (3 and 4) and consider the need determination processes so far excluded; for these processes will almost certainly lead previously undifferentiated users (equivalent to the assumption of "identical users") to reveal differentiated behavior; that is, submarkets will emerge.

User differences may eventually appear for a variety of reasons: different uses, differential user adaptation to products, differential preferences with respect to price versus quality, and differential attitudes to risk.[18] Not all of these differences are obvious at the outset of the process. For example, the effects of different uses on the definition of user needs may require a period of experiencing the various options available, while the effects of differential preferences

---

18. The last two factors can be considered together with or independently of differences in income.

with respect to price versus quality or of differential attitudes to risk may only become felt as the result of a string of new product-price options appearing in the market, that is, as the result of a dynamic technology. We shall attempt to describe the conditions most likely to generate each one of the above reasons for user differences.

*Differential user adaptation to products.* Assume undifferentiated users initially and a random initial option chosen by each of them. If technology is nonmalleable, then submarkets may appear. Since the redesigning of products involves cost and is not instantaneous, a certain amount of user adaptation must occur. If it is sufficiently strong and product-specific, it will generate separate need-determination paths.

*Different preferences concerning price versus quality.* These differences may arise from a dynamic technology, such as new process technology leading to lower-cost standard products (like Ford's Model T). Users previously revealing undifferentiated behavior may now divide themselves into two groups: those choosing the new low-price/low-quality option and those choosing the previous high-price/high-quality option. These differences may (but need not) reflect different income levels, but income is relevant only for distinguishing users after the new product-price option has appeared. Similarly a product improvement may also lead to user differentiation: between those who will eventually consider the new factors as essential or relevant and those who will not.

*Different attitudes to risk.* These may be revealed by analyzing user reaction to a major product improvement, e.g., the application of laser technology to surgical equipment. Under these conditions, the productivity of the improvement is uncertain, at least in the early stages, and only a subgroup of users — less risk-avoiding than the rest — might shift to the new option. A new process of need determination and a corresponding submarket will emerge.

*Different uses.* Some differences in uses will be obvious from the outset, but some may be revealed only through product innovations. In the field of nuclear medicine, the new data-processing and display possibilities appearing in the early 1970s have separated those laboratories studying the functioning of organs (e.g., those specializing in cardiology), and therefore requiring gamma cameras incorporating the new features, from those laboratories engaged in other types of research and not requiring this newer technology (e.g., laboratories specializing in the imaging of bones). The differentiation of uses has partly resulted from the new feature, and this illustrates how a dynamic product technology may certainly lead to the emergence of submarkets based on different uses.

Our first conclusion is that a dynamic technology may, while stimulating need determination, also lead to the emergence of submarkets. The process can be represented as in figure 4.3$a$, which shows a hypothetical example referring to computers. Suppose that $d_1$ and $d_2$ are (respectively) effective storage capacity and computing speed, and $d_{1b}^*$ and $d_{2s}^*$ are the minimum requirements for business users ($b$) and scientific users ($s$). There are two first-generation product options, $v$ and $w$, whose location in ($d_1$, $d_2$) is uncertain and may be anywhere within the shaded area; neither user has a definite preference for one or the other option (we ignore differences in price). Two improvements, $A$ and $B$, with definite locations in ($d_1$, $d_2$) now make their appearance. Improvement $A$ is preferred by business users and $B$ by scientific users. Under the assumption that $v$, $w$, $A$, and $B$ are each associated with specific product features, improvements $A$ and $B$ have induced need determination, and two distinct submarkets have emerged.

The number of submarkets may, however, decrease once a new product-price option appears. This will be particularly true with radical new-product innovations such as the appearance of a completely new product class relevant to the broad need under consideration or of a new generation of equipment. This possibility is shown in figure 4.3$b$. Against the background of two clearly defined submarkets, $b$ and $s$, a new generation of equipment, comprising product options $x$ and $y$, makes its appearance. The location of $x$ and $y$ in ($d_1$, $d_2$) is uncertain and can be anywhere in the shaded region. Assume that the innovation has been sufficiently radical to ensure that any point in the shaded region is preferable to both $A$ and $B$. The result will be that both business and scientific users will tend to shift to the new generation of equipment despite the fact that neither may be able to express a definite preference for the specific product features of $x$ and $y$. Experience, adaptation, and minor improvements will then again induce need determination and the appearance of submarkets.

### Implications for product standardization and competition

The preceding discussion is consistent with the view that a dynamic technology will be associated with a complex pattern of product change involving two main elements: (1) within a given product class (or fundamental principle of operation), the continual appearance of a generally increasing set of increasingly well-defined products which replace those with more or less random product differences; (2) the discrete appearance of new product classes (or generations) which replace existing ones. The well-defined products in (1) correspond to the submarkets generated by need determination in the context of nonradical improvements, with product standardization occurring *within* each submarket. The second element consists of radical product innovations which set off a new round of need determination and submarket generation as in (1).

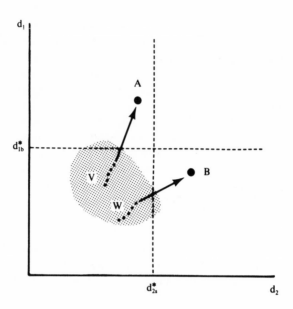

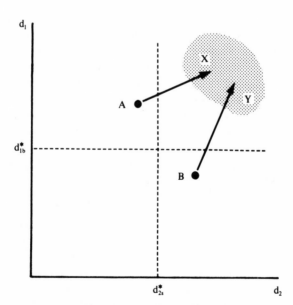

Fig. 4.3. Submarket formation and elimination.
(a) effects of minor improvement: need determination and the
emergence of submarkets. (b) effects of major improvement: elimination of submarkets.

The emergence of a new product class represents a break from the immediate past whose immediate manifestation could be a (temporary) reduction in the number of submarkets[19] and temporary reversal of need determination.

Each one of the above elements of product change contradicts the simple product-cycle view of increasing standardization. The strongest contradiction comes from the appearance of new product classes or new generations, and the critical issue is the frequency of this phenomenon. Part of the answer should be sought in the origin or sources of the technological knowledge impinging on the cycle considered. When this knowledge is increasingly generated by firms or institutions active in other areas, we might expect an increasingly varied set of fruitful combinations of knowledge which would provide a permanent potential for significant product innovations. Such cross-fertilization seems to be characteristic of an increasing number of cycles in modern industrial society.[20] A necessary condition for it to occur is a technology-oriented firm strategy, i.e., a strategy in which the firm plays an *active* role in launching new product options and is not satisfied with responding to the identification, specification, and translation of user needs existing at the time.

The emergence of well-defined submarkets also contradicts the simple view of increasing product standardization. A major factor limiting the number of submarkets is the number of significant process innovations introduced, and this in turn will depend on the potential for exploiting economies of scale in production. Size of market (actual and potential) will be a significant variable here.[21]

The incentives for product competition may also be enhanced under our pattern of product change. On the one hand, the emergence of a new product class opens up a wide range of new technical opportunities for adapting it to existing and new uses; on the other, need determination and the gradual emergence of submarkets will reduce the innovation risks associated with the understanding of user needs and with product acceptance in a variety of settings, thereby providing additional stimulus to competition based on product innovation. These incentives may persist up to the point of exhaustion of the innovation possibilities associated with the new product class in its various uses. Price competition may then become increasingly important,[22] but the stage may be set for the rapid appearance of a new product class.

19. At least for innovations which command wide acceptance even in the short run.

20. This is the essence of Walker's (1975) argument against product standardization. He asserts that within a significant segment of modern industry new generations of equipment follow each other too rapidly, although standardization does occur within a particular generation.

21. The potential for exploiting economies of scale in production would be higher in, say, automobiles than in scientific instruments.

22. As emphasized by Vernon (1966), Hirsch (1967), and Utterback and Abernathy (1975).

*Cycle types*

We now present a preliminary classification of cycles which may lead to some useful implications.

*Basic substances* are defined as perfectly divisible commodities whose relevant characteristics are expressed exclusively in terms of chemical-physical structure.[23] It follows that product adaptation to user needs does not include a price element and that the process of need determination does not include either an identification stage or a stage of translation of user needs. The process of need determination would then involve the use or testing of alternative chemical compounds until one is found with the optimum chemical-physical structure.

The properties of basic substances differ from the properties of the commodities underlying the analysis of this paper, which we term *systems*. These differences lie at the root of the distinction between the two groups of commodities with respect to the type of competition prevailing and to the factors giving competitive advantage to firms or countries. We may summarize some of these differences as follows: (1) the scope for product competition is likely to be narrower in basic substances than in systems; (2) marketing strength and marketing effort are likely to be less important relative to R & D strength and R & D effort for success in (basic) substance innovation than for success in systems innovations.

We briefly discuss the reasons for these conclusions. Basic substances possess fewer product dimensions that can be varied since the product dimensions all relate to chemical-physical structure rather than to various degrees of specificity (functions, features, etc.), as is the case with systems. This does not necessarily imply that product competition is not important; it may be so, but only during the period when user needs have not been perfectly defined, or after that, if the nonmalleable character of the technology prevents all firms from obtaining the optimum chemical-physical structure. After this, price competition will prevail. In contrast, product competition in systems will still be important beyond the specification of user needs, during the stage of translation into terms of concrete characteristics. Moreover, during this stage the number of product dimensions that may be varied is very large.[24]

23. Thus sulfuric acid but not sugar can be regarded as a basic substance: sugar would be only a substance, since its relevant characteristic would be sweetness and not a particular chemical-physical structure.

24. Another reason product competition is more dominant in systems is that it will prevail before identification, whereas our definition of basic substance excludes the need of identifying a product class. However, competition between different substances might prevail before any substance has become basic. This is the reason why the class of basic substances, while well differentiated from systems, may be comparatively unimportant in the real world.

The implication of the above is that factors which reduce the production costs of basic substances are likely to be critical for the competitiveness of the firms and countries producing them. These factors include R & D resources for process innovations, physical capital to take advantage of economies of scale in production, and low wages. Marketing resources, the size and sophistication of local markets, and proximity to world markets will, on the other hand, be relatively more important for competitiveness in systems. The existence of ample R & D resources will be as critical for systems as they are for basic substances since they will make possible efficient translation of perceived user needs into new products (i.e., for product innovations).

### Conclusions

Our original objective was to develop the notions of user needs and need determinateness in a way which would be useful for explaining the commercial performance of innovations. It became necessary to explain the process of need determination, i.e., of how user preferences get to be defined more and more specifically in terms of product class, performance dimensions, and features. It rapidly became clear that conditions of supply play a very important role in this process and in particular the appearance of a new product-price option might accelerate it considerably.

We thus find ourselves squarely in product-cycle theory, a central theme of which is the process of product standardization. Although need determination may, under very restrictive assumptions, lead to product standardization, it will in general—particularly if induced by product improvements—lead to the appearance of distinct submarkets. In other words, users with undefined preferences, and therefore undifferentiated behavior, progressively get classified into one of several groups, each one revealing preferences that are sufficiently defined to be different from the others. This modifies the conventional product-cycle view of product standardization and opens up additional opportunities for product competition (resulting from a reduction in the risk that the product is not adapted to user needs or not perceived as such). Another implication is that it may be very difficult to characterize differences in user preferences a priori on the basis of a given set of parameters, since these differences reveal themselves in the course of time as a function of the stream of product-price options appearing on the scene.

The main deficiency in the analysis lies in its treatment of the supply side of the cycle. The innovations which affect need determination are basically exogenous to the analysis although it is true that higher need determinateness has been shown to reduce some risks associated with innovations. The overall innovation strategy of the firm and in particular the matching of a subset of feasible innovations with a subset of marketable ones have not been considered at all.

## Appendix: Need Determinateness and Quality Evaluation

There are two sorts of inputs which enter the quality evaluation of products at any particular time: user experience up to $t$ and information-gathering activities performed at $t$. User experience with products up to $t$ is reflected by the extent of need determinateness attained at $t$. Information-gathering activities involve both the inspection of product characteristics and the testing of products or components. Here we use the term testing to refer exclusively to actual use of the product before purchase in order to ascertain its quality directly. Although user experience is always likely to be relevant for evaluating quality, its importance is enhanced if gathering information is costly or inefficient or both. For example, there are some products such as complex systems or machines which cannot be reliably or even safely tested, so that the only information obtainable concerns product characteristics.[25] For this class of goods ("experience" goods in Nelson's [1970] terminology) quality evaluation depends primarily on need determinateness, although inspection will also play a role. We present below a mechanism for quality evaluation of these goods which also will explain the relationship between need determinateness and uncertainty. Our specific assumptions concerning information-gathering activities are as follows:

> A1. Testing of products is either not feasible or if feasible it is not worthwhile because of high cost or unreliability or hazard, or all three.
>
> A2. Inspection is carried out only with respect to specifications and is costless.

Assumption A2 implies that more general product characteristics such as features and functions are not ascertained by inspection but by other means. Under A1 and A2 the process of quality determination consists of stages:

| Determining product specifications | → | Product assessment | → | Quality evaluation |
|---|---|---|---|---|

Product assessment refers to the less specific aspects of a product, such as features or functions, which are not as easily checked as product specifications. These product aspects are inferred from product specifications with the object of determining performance dimension levels. Once user needs have been determined, this is all the information required for quality evaluation.

We therefore concentrate here on the process of product assessment. This may be more or less specific; e.g., assessment of product features is more specific than assessment of product functions. We postulate the following propositions:

---

25. The class of products for which this is significant may be wider than one might think since the future quality of durable goods can never be ascertained by current testing.

P1. The specificity of the product assessment required for quality evaluation varies directly with the degree of need determinateness.

P2. The uncertainty associated with a particular product assessment varies inversely with its specificity.

P1 implies that when need determinateness is low, relatively general aspects of the product, such as performance dimensions, have to be inferred directly from the specifications. On the other hand, if need determinateness is high, it may be sufficient to assess product features in order to evaluate quality. This requirement follows from the definition of need determinateness. P2 implies that there is greater uncertainty in assessing product functions than product features from specifications. As well as the uncertainty in relating features to specifications there is the additional uncertainty of deriving functions from features.

A consequence of this is our third proposition:

P3. Product quality evaluation is less uncertain, the greater the degree of need determinateness attained.

The relationship implied in P1–P3 can be summarized as follows:

| Degree of need determinateness | → | Specificity of product assessment required | → | Certainty of quality evaluation |

### Relation to previous work

There are at least three related issues on information and quality which touch upon the issue raised in this paper: (1) determining the amount of information collected or, alternatively, the intensity of information-gathering activities, (2) determining the kind of information-gathering activities (or the procedures for obtaining information) to be used in different circumstances, (3) postulating mechanisms for evaluating quality. The first of these has been dealt with by several authors, including Nelson (1970). His aim was to analyze the implications of the different amounts of information on brands gathered for two classes of goods (search goods and experience goods) for industry concentration. These classes of goods correspond to the two alternative procedures for obtaining information on product quality that he postulates: search, which corresponds roughly with what we have called inspection and testing; and experience, by means of purchase. He postulates a probability distribution of objective quality among brands and perfect certainty in evaluating the relative quality of the brands on which information has been collected. Thus there exists ex ante uncertainty about whether the next considered will turn out to be of high quality or low quality, but there is *no* ex post uncertainty about quality or the quality ranking of goods once information about the brand has been collected. In neither case does he postulate any *mechanism* for evaluating product quality.

In this paper we concentrate precisely on this mechanism since the uncertainties involved have been found to be important in explaining innovation performance. Our purpose is *not* to explain the amount of information gathered by consumers on alternative options but rather to explain how user uncertainty about the quality of products declines as need determinateness increases.

This issue and the associated mechanism for quality evaluation would also be relevant in Nelson's analysis once it is recognized that his method of building up a sample for durables is irrelevant, especially in the face of the continuing introduction of new models.

### References

Hirsch, Seev. 1967. *Location of Industry and International Competitiveness.* Oxford: Clarendon Press.

Ironmonger, D. S. 1972. *New Commodities and Consumer Behavior.* Department of Applied Economics, Monograph 20. Cambridge: Cambridge University Press.

Jones, H. Ralph. 1969. *Constraints to the Development and Marketing of Medical Electronic Equipment.* Ann Arbor: Institute of Science and Technology, University of Michigan.

Lancaster, Kelvin. 1971. *Consumer Demand: A New Approach.* Columbia Studies in Economics, no. 5. New York and London: Columbia University Press.

Nelson, Phillip. 1970. "Information and Consumer Behavior." *Journal of Political Economy* 78 (March/April): 311–29.

Science Policy Research Unit, University of Sussex. 1972. *Success and Failure in Industrial Innovation: Report on Project SAPPHO.* London: Centre for the Study of Industrial Innovation.

Sweeney, James L. 1974. "Quality, Commodity Hierarchies, and Housing Markets." *Econometrica* 42 (January): 147–67.

Teubal, Morris, Naftali Arnon, and Manuel Trachtenberg. 1976. "Performance in Innovation in the Israeli Electronics Industry: A Case Study of Biomedical Electronics Instrumentation." *Research Policy* 5: 354–79. (Chap. 1 of this volume.)

Utterback, J., and W. Abernathy. 1975. "A Dynamic Model of Process and Product Innovation." *Omega* 3: 639–56.

Vernon, Raymond. 1966. "International Investment and International Trade in the Product Cycle." *Quarterly Journal of Economics,* 80 (May): 190–207.

Walker, W. S. 1975. *Industrial Innovation and International Trading Performance.* Science Policy Research Unit, University of Sussex.

Many of the ideas appearing here were developed jointly with Manuel Trachtenberg. I appreciate the comments on previous drafts by S. Freund, R. Nelson, and M. Trachtenberg and the discussions with R. Gronau, G. Hanoch, R. Klinov-Malul, M. Spechler, and P. T. Spiller. I am grateful to the National Council for Research and Development, the Prime Minister's Office, Jerusalem, for financial help.

# 2    **Learning, Technological Capabilities, and Spin-Offs**

# Introduction to Part 2

The importance of technological development in industrialization is gradually being recognized by economists. An increasing number of studies show that despite the lack of original invention in less developed countries, the incorporation of an increasing array of new products and processes inevitably implies that the country has at least assimilated foreign technology. (See Katz 1980; Westphal, Rhee, and Pursell 1981; Westphal, Kim, and Dahlman 1984; Dahlman and Valadares Fonseca 1978; Kim 1980; Maxwell 1976.) Moreover, this assimilation is not instantaneous or costless. It generally requires explicit investments of various kinds (in knowledge, human and physical capital) and organizational change. Learning of various kinds is involved, and judicious government policy directed to the acquisition of technological capabilities may make an important difference. Chapter 5 studies the impact of technological learning and other spin-off mechanisms on the exports of selected Brazilian capital goods. The paper, which is based on in-depth interviews of eight important capital-goods producers, argues that the significant increase in Brazilian capital-goods exports during the seventies and in particular the rise in the share of output exported cannot be explained as effects of the direct short-term foreign trade instruments used (tariffs, exchange rates, export subsidies) since these have not affected the relative profitability of the export and local markets for capital goods. Other measures or factors were responsible, such as the firm learning processes considered in the paper. A basic distinction made in the paper is between the acquisition of manufacturing capabilities and the acquisition of a design capability. The former refers to the gradual mastering of an increasing number of manufacturing processes (such as machining, welding, and assembly) and the increased knowledge and experience in the use of materials. A design capability exists when the firm can also specify the capital good required by the user, which requires a deeper knowledge of materials and knowledge of the production process into which the capital good will fit.

A design capability, however, need not involve R & D and prototype building and testing. In many cases it merely tailors a product to a particular application without modifying the basic product design.

The contribution of early activities to the later capabilities of the firms studied is not due only to learning; there are other spin-off mechanisms involved as well. Of the 11 different ones that were identified and that had at least some relevance for exports, four mechanisms were related to manufacturing and three to design. The remaining four mechanisms include the extremely important firm reputation effect: in some instances, the activity related to the local market significantly enhanced subsequent export potential as a result of this effect (e.g., in enabling Brazilian firms to prequalify for international tenders). The most common spin-offs identified related to manufacturing (e.g., more precise machining abilities). They were present in every firm but one and were extremely significant; and the associated capabilities generally became established before the emergence of design capabilities (with the exception of the engineering consulting firm of the sample). Design capabilities, if emerging at all, relate to a narrower set of products. Sometimes, the link between the two types of capabilities is composed of the procedures, equipment, and personnel involved in quality control (e.g., in the firm designing and producing pistons and bearings for the automobile market). Finally, the study suggests an important distinction between simple, or process, learning and product learning. The former enables the firm to produce existing equipment at lower cost (a learning *curve* effect) or to introduce relatively minor technological improvements. The latter enables it to shift to new, but in some respects technologically related, types of equipment which are more difficult to produce or which require a design capability as well. Product learning has enabled the production of goods with more exacting specifications (e.g., pipes, pistons) and of larger size (e.g., mixers for the pharmaceutical, food, and nuclear reactor industries). Firm capabilities have also evolved from simple product designs to a capability of offering complete packages (e.g., turnkey alcohol plants). In all cases except one, spin-offs involving product learning have led to exports.

### The Classification of Technological Capabilities

Westphal, Rhee, and Pursell (1981) propose three broad categories of capabilities (or associated learning processes) related to industrialization: learning to produce, to invest, and to innovate. Learning to produce includes the operation of whole plants as well as the skills associated with individual machining operations. Learning to invest refers to the identification, design, and execution of investment projects, including the start-up of newly erected production facilities. Finally, innovation capabilities enable R & D, via product prototypes or experimental (process) plants, to successfully launch new products or processes. The literature on industrialization and technology transfer has emphasized the first two categories: production and investment capabilities. The Brazilian capital-goods study has identified mechanisms of learning to produce and some aspects of learning to innovate—although many of the design capabilities identified (i.e., those associated with the "tailoring" of capital goods) do not reflect innovation capabilities per se. They would more appropriately be classified with production capabilities. One instance only of accretion of investment capabilities has been identified—in relation to pipe plants. Presumably a much larger incidence of learning to invest mechanisms would have been found if our study of capabilities would have focused on process industries or steel (e.g., see Dahlman and Valadares Fonseca 1978; Maxwell 1976, 1982; and Westphal, Rhee, and Pursell 1981). An important aspect of the Brazilian study is, I believe, its emphasis on qualitative, or product, learning, i.e., learning which enables the firm to produce, invest, and even design increasingly complex products or systems.

There are at least two other learning mechanisms mentioned in the literature which are worthwhile mentioning here: Rosenberg's "learning by using" (Rosenberg 1982, chap. 6) and what may be termed user-producer interactions. The former relates, for example, to the experience and knowledge accumulated from operating or using an industrial plant or an aircraft. This learning may increase any one of the capabilities mentioned above: production (or operation of a process), an increase in which can lead to the phenomenon called capacity stretching (see Dahlman and Valadares Fonseca 1978; Maxwell 1976; and Maxwell and Teubal 1981); investment (an example of this in the Brazilian study is Confab's experience with operating its first pipe plant, which improved its capability for designing an efficient subsequent plant); or design (Rosenberg's designs of "stretched aircraft" may be a case in point). Learning from user-producer interactions relates first and foremost to innovation capabilities, i.e., the active contribution of users in the innovation process of firms, especially those active in the machinery and instrumentation areas (see von Hippel 1976, 1978; Pavitt 1984).

### Spin-Offs and Innovation Capabilities

Existing literature has given very little prominence to the learning mechanisms associated with innovation, and this is because industrialization generally begins with industries which are not particularly R & D intensive. In the development of steel, automobiles, shipbuilding, and capital-goods industries, the initial technological issues refer to the acquisition of production and investment capabilities. This is not so for the electronics industry, especially the R & D intensive segment of it: R & D and associated capabilities exist or emerge at the outset. Chapter 6 analyzes the R & D performance through time of a successful Israeli electronic instruments producer. The R & D history was expressed in terms of a series of R & D projects representing successive generations of a product, transitions to more complex substitute products, and diversification toward other products. Qualitative information on the various projects suggested strongly that the profitability of a project at time $t$ depended considerably on the projects undertaken prior to $t$; i.e., considerable spin-offs flowed from early projects and benefited later ones. The nature of the spin-offs and learning processes of this electronics firm seem to differ considerably from those shown in the learning and capabilities development literature. An important category of spin-offs in the electronics industry is that related to R & D, both specific technological knowledge which may benefit future design work *and* overall experience in designing instruments adapted to the market place. Another important category relates to marketing, user feedback, and a product-line effect. As with capital goods, reputation effects are of considerable importance. An important highlight of the paper is the attempt to *quantify* the economic value of (some of) the spin-offs generated during the growth of the firm. The share of spin-offs in the *total* (in contrast to the *direct*) profitability of early projects was very significant. Current research on an Israeli communications firm confirms the importance of these and other spin-offs related both to R & D and to marketing. Spin-offs related to production were also identified, although they seem to be of lesser importance relative to their role in the metalworking and capital-goods firms previously analyzed in the literature. In this connection, chapter 7 of this volume undertakes a systematic comparison of the types of intangibles (spin-offs) accumulated in three types of industries (based on available case studies): operating experience and engineering capabilities in the process industries; manufacturing abilities in metalworking; and a variety of intangibles (especially from R & D) in electronics. The paper shows that the intersectorial differences in learning are significant and that they go beyond the obvious differences due to product cycle considerations. This could be extremely important for policy, especially for infant industry promotion based on enhancing the accretion of capabilities.

Chapters 11 and 12 claim that one of the effects (justifications) of an industrial R & D promotion policy in Israel has been (is) to generate what may be termed an *industrial innovation capability*, i.e., a capability to commercially launch new products and processes in industry. The case study reported in chapter 6 and the study of industrial R & D support in Israel (chap. 11) give some concreteness to the meaning of such a capability and to the process by which it may be promoted (Nelson's [1982] work provides a theoretical basis for this capability). This may be relevant for those newly industrialized countries wishing to diversify into electronics.

## References

Dahlman, C. J., and F. Valadares Fonseca. 1978. "From Technological Dependence to Technological Development: The Case of the Usiminas Steel Plant in Brazil." ECLA/IDB/IDRC/UNDP Research Programme on Scientific and Technological Development in Latin America (hereafter abbreviated Res. Prog. Sci. Tech. Dev. Latin America), Working Paper no. 21. Buenos Aires.

Katz, J. 1980. "Domestic Technology Generation in LDC's: A Review of Research Findings." Res. Prog. Sci. Tech. Dev. Latin America, Working Paper no. 35. Buenos Aires.

Kim, L. 1980. "Stages of Development of Industrial Technology in a Developing Country: A Model." *Research Policy* 9:254–77.

Maxwell, P. 1976. "Learning and Technical Changes in the Steel Plant of Acindar S.A. in Rosario, Argentina." Res. Prog. Sci. Tech. Dev. Latin America, Working Paper no. 4. Buenos Aires.

Maxwell, p. 1982. "Steel Plant Technological Development in Latin America." Res. Prog. Sci. Tech. Dev. Latin America, Working Paper no. 55. Buenos Aires.

Maxwell, P., and M. Teubal. 1981. "Capacity Stretching Technical Change: Some Theoretical Aspects." The Maurice Falk Institute for Economic Research in Israel Discussion Paper no. 815. Jerusalem.

Nelson, R. 1982. "The Role of Knowledge in R & D Efficiency." *Quarterly Journal of Economics* 447:453–70.

Pavitt, K. 1984. "Sectoral Patterns of Technical Change: Towards a Taxonomy and Theory." *Research Policy* 13:343–73.

Rosenberg, N. 1982. *Inside the Black Box: Technology and Economics.* Cambridge: Cambridge University Press.

von Hippel, E. 1976. "The Dominant Role of Users in the Scientific Instrument Innovation Process." *Research Policy* 5:212–39.

von Hippel, E. 1978. "A Customer-Active Paradigm for Industrial Product Idea Generation." *Research Policy* 7:240–66.

Westphal, L., L. Kim, and C. Dahlman. 1984. "Reflections on Korea's Acquisition of Technological Capability." Washington, D.C.: World Bank.

Westphal, L., Y. Rhee, and G. Pursell. 1981. "Korean Industrial Competence: Where It Came From." World Bank Staff Working Paper no. 469. Washington, D.C.

# 5 The Role of Technological Learning in the Exports of Manufactured Goods
## The Case of Selected Capital Goods in Brazil

## Morris Teubal

### Introduction

The object of the study is to explain the evolution of Brazilian capital-goods exports in terms of at least two sets of variables: technological learning within the sector; and government policy.[1] The learning variable is emphasized at the expense of more traditional economic variables, such as size of sector (or level of activity) and government policy variables, for several reasons. Most studies of exports put emphasis on these traditional variables;[2] although intuitively it is of significance for exports of nonhomogeneous products, the technological learning variable has received very little attention from economists; it is also a critical variable for understanding the development of infant industries and the emergence of dynamic comparative advantage, and its conceptualization will therefore have policy implications. However, the study does not concentrate on technological learning to the exclusion of government policy variables. While the direct effects of the latter on exports are pretty well known, the indirect effects, via the effect on learning, have not been investigated.[3] This paper attempts to illustrate and illuminate the nature of these relationships.

The approach followed is to identify relevant learning processes for eventual introduction into export-behavior equations or as a basis for theoretical models.

1. A third important variable is the environment facing the sector, principally the domestic and export markets. Some of these will be referred to indirectly; e.g., a principal component in the domestic market for capital goods was equipment demanded by state-owned enterprises (a government policy variable).

2. An example is Tyler's 1980 study on Brazilian exports. An important exception is Katz and Ablin 1976.

3. Some policy instruments, such as exchange rates and subsidies, directly affect the share of output which is exported. Others, such as output subsidies and protection of the domestic market, may have only an indirect effect on exports, e.g., via learning and exploitation of economies of scale. Econometric work on exports has emphasized exchange rates and subsidies.

Its starting point was a series of microeconomic case studies of technologically sophisticated firms (eight firms were interviewed in Brazil). Such studies concentrate on a small number of variables focusing on the learning processes of firms, and in particular on the indirect contribution (spin-offs) of early activities or products to later activities or products. Among the specific issues addressed in the case studies are (1) the nature of technological learning (are there recognizable sequences and does this new knowledge generally lead to products of increased technological complexity or sophistication?); (2) how (if at all) technological learning has benefited exports; (3) other endogenous firm spin-offs, principally from reputation effects, which were important for increased efficiency and for the emergence of exports; (4) the economic value of the firm's accumulated experience, and the factors determining it.

### The learning process

The learning process within the firm relates to a whole series of activities: manufacture and plant operation; investment or project execution; product design and R & D. The case studies document all aspects of capacitation but some of those more commonly found should be emphasized here. A basic distinction, which is particularly relevant for metalworking industries, is between learning to manufacture or the acquisition of manufacturing capabilities and the *acquisition of design capability*. The former refers to the gradual mastery of an increasing range of manufacturing processes, such as machining and other metal-forming processes, welding, assembly. It also involves increasing knowledge of and experience with the many materials used. The acquisition of design capability enables the firm to specify the equipment (product) required for a particular manufacturing process (user). This requires a much deeper knowledge of materials in order to be able to select the best for each purpose. It also requires knowledge about the use to which the product will be put. For example, the manufacturer of chemical-industry equipment must be familiar with the entire process carried out by the customer, not merely that part of it in which his equipment is involved. It is customary to distinguish detail design capability from basic design capability The former tailors a product to a particular application without modifying the general type or class. Basic design capability, on the other hand, may enable a firm to adapt existing product types or to launch completely new products (innovation). R & D is the set of activities (such as the construction and testing of prototypes or pilot plants) which, in firms that have basic design capability, leads to product or process innovations.

Learning may or may not lead to product or process adaptation, depending on a number of factors. Manufacturing abilities alone are probably insufficient to sustain product or process adaptation, although minor adaptations may be possible. On the other hand, the acquisition of design capability does not by itself ensure that more than minor adaptations (and tailoring) take place. This depends on other factors as well, some of them related to the market. However, the presence of design capability increases the probability of major adaptations.

### Why capital goods?

The focus is on capital goods in Brazil. The main reason for choosing these industries in the first place was their impressive development, especially during the "Brazilian miracle" period (1968–1974), not only quantitatively but qualitatively as well. The array of goods being produced and exported and their increasing sophistication are prima facie evidence of intensive technological learning and identifiable adaptation. In addition, capital goods, broadly defined to include all transport equipment, have in the last decade become the most important component of exports of relatively sophisticated products.[4] Other a priori considerations (some of them not specific to capital goods) were (1) machinery or metal products are nonhomogeneous products: the large number of different features increases the probability of product adaptations; (2) process adaptations in other industries (such as raw-materials processing) may involve machinery adapted by the capital-goods sector; (3) machinery production may be divided into a set of elementary tasks common to a wide range of machinery types (see Rosenberg 1976, chap. 1). Thus the probability of learning and adaptation would seem to be high. Needless to say, I do not claim that all the relevant learning processes are concentrated in the capital-goods sector (although this may in part depend on what is classified as capital goods).[5]

The next section gives an overview of the development of the capital-goods sector and its exports. It is shown that the increased share of capital-goods exports is part of a general increase in the sophistication of total Brazilian exports. The work of others on the effects of government policy on the evolution of exports is summarized in the third section. It is argued that direct short-term foreign-trade instruments (tariffs, exchange rates, export subsidies)

---

4. See Dahlman's report (1981, 40); "relatively sophisticated" products were defined as manufactured products (International Standard Industrial Classification definition) excluding semiprocessed basic products such as brown sugar, frozen meat, and processed mineral ores; semimanufactured products such as crystallized sugar, natural wax, vegetable oils, cut woods, cacao paste, processed hides, paper pulp, and iron and steel in crude form and in simple products; other products using relatively simple technology such as food products, beverages, textiles, clothing, leather goods, footwear, and wood products. The share of capital goods in total exports of relatively sophisticated products rose from 30% in 1970 to around 50% in 1979.

5. Points 1 and 3 are common to metal products. Also, some of the factors favoring product adaptations in nonhomogeneous products may favor process adaptations in the chemical and petrochemical industries.

cannot explain the rise in the share of capital-goods output exported, since they do not seem to have affected the relative profitability of the export and local markets for capital goods. This points to other measures and factors, such as increased efficiency in the sector (for which there is independent evidence). One cause of this increased efficiency—the learning processes within the firm—is investigated in the fourth section. The last section summarizes the paper and spells out some implications of the analysis for productivity (and export) growth and for the theory of the infant firm. It should be said at the outset that the present work does not lead to a clear-cut theoretical model of learning and export growth, nor to a quantitative empirical relationship between them.

## Capital-Goods Exports: An Overview of the Brazilian Case

The growth of Brazilian capital-goods exports should be related to the growth of manufactured-goods exports in general and to the growth of relatively sophisticated manufactured exports, of which capital goods is an important group. Table 5.1 summarizes table 1 of the Dahlman report (1981). The first two

Table 5.1. Development of Brazilian exports: 1970–1980

| | Total exports | | Manufactured products | | Industrialized products (Brazilian definition) (5) | Relatively sophisticated products[a] (based on ISIC) (6) | Capital goods[b] (7) |
|---|---|---|---|---|---|---|---|
| | Millions of current U.S. $ (1) | Millions of 1970 U.S. $ (2) | ISIC definition (3) | Brazilian definition (4) | | | |
| 1970 | 2,739 | 2,739 | 70.4 | 15.2 | 24.3 | 10.9 | 3.4 |
| 1971 | 2,904 | 2,792 | 69.9 | 20.0 | 28.3 | 11.4 | 4.3 |
| 1972 | 3,991 | 3,695 | 76.6 | 22.9 | 30.6 | 12.7 | 4.9 |
| 1973 | 6,199 | 5,347 | 72.8 | 23.6 | 31.3 | 9.3 | 3.6 |
| 1974 | 7,951 | 6,260 | 70.4 | 28.5 | 40.0 | 12.9 | 6.8 |
| 1975 | 8,670 | 6,282 | 66.2 | 29.8 | 39.6 | 19.1 | 9.0 |
| 1976 | 10,128 | 6,889 | 69.3 | 27.4 | 35.7 | 16.1 | 8.2 |
| 1977 | 12,120 | 7,769 | 74.4 | 31.7 | 40.3 | 18.6 | 10.1 |
| 1978 | 12,659 | 7,535 | 79.4 | 40.2 | 51.4 | 25.0 | 13.4 |
| 1979 | 15,244 | 8,195 | 80.0 | 43.6 | 56.0 | 31.1 | 14.7 |
| 1980 | 20,132 | — | — | 44.9 | 56.5 | — | — |
| 1970–1979 | 5.6 | 3.0 | 1.1 | 2.9 | 2.3 | 2.9 | 4.3 |

Percentage of total exports

Source: Dahlman 1981, table 1.

[a]The category of relatively sophisticated products includes—in addition to capital goods—paper and paper products, printing and publishing; industrial and other chemicals; petroleum refining, rubber and plastic products, basic ferrous and nonferrous metal products, etc.

[b]Includes motor vehicles.

columns show the development of total Brazilian exports between 1970 and 1980. Nominal exports in 1980 were 7.3 times the 1970 level, while real exports trebled between 1970 and 1979, an impressive record by any standard. The shares of various definitions of manufactured or industrialized products in total exports are shown in columns (3), (4), and (5). The ISIC definition of manufactures includes anything which is manufactured and thus includes processed food, most of which relies on relatively simple technology. The Brazilian definition of industrialized products excludes a number of semiprocessed products, while the Brazilian definition of manufactured products subtracts semimanufactures from industrialized products (see n.4 above). The figures of these three columns show that while the share of manufactures (ISIC) in total exports has risen by about 10 percentage points, the share of the industrialized and manufactured products (Brazilian definition) doubled and trebled, respectively. The increasing sophistication of Brazil's exports stands out when we look at the share of the last two categories in table 5.1. Capital goods (column [7]) are an important component of relatively sophisticated exports, and their share in total exports quadrupled between 1970 and 1979. Thus the increasing importance of capital-goods exports can be viewed as part of a wider trend of increasing technological sophistication. This is also true of the output of capital goods.

### Production and exports

The central feature of Brazilian capital-goods[6] exports during the last fifteen years is their impressive growth, both absolutely and relatively to the output of the sector. From $87 million in 1970, representing 2.2% of output, exports grew to $800 million (1970 dollars) in 1978, or 9% of the output of that year.[7] Real exports grew at an annual rate of 28% during the period. The growth in the share of output exported took place in a sector whose real annual growth rate (11%) exceeded that of the manufacturing industry in general—the share of capital goods in total manufacturing output increased from 11.5% in 1965 to 15.1% in 1977 (World Bank 1980, 65).

The composition of capital-goods exports, including automobiles, between 1970 and 1979 is seen in table 5.2. The largest categories are transport and nonelectrical machinery, followed by electrical machinery, metal products, and

6. Unless otherwise stated, the definition of capital goods used is that appearing in World Bank 1980. It includes metal products, mechanical and electrical equipment, and transport equipment other than automobiles.

7. See World Bank 1980, 67. Exports of a broader category that includes transport equipment, boilers and machinery, electrical equipment, and steel manufacture grew from $213 million in 1972 to $1920 million in 1978 and $2595 million in 1979 (see World Bank 1981). A summary of some of the information can be found in table 5.2.

Table 5.2. Production and exports of capital goods: 1970–1979

| | Total[a] (millions of 1970 U.S. $) | | Exports[b] (current prices) | | | | | |
|---|---|---|---|---|---|---|---|---|
| | Output | Exports | Total | Metal products | Nonelectrical machinery | Electrical machinery | Transport equipment | Scientific and other equipment |
| Millions of U.S. $ | | | | | | | | |
| 1970 | 3,203 | 87 | 105.7 | 9.8 | 63.6 | 16.5 | 14.9 | 0.9 |
| 1971 | 3,610 | 114 | 148.8 | 11.5 | 76.2 | 27.9 | 30.8 | 2.4 |
| 1972 | 4,366 | 182 | 239.7 | 18.8 | 98.4 | 38.7 | 80.9 | 2.9 |
| 1973 | 5,920 | 241 | 336.1 | 25.0 | 128.9 | 83.4 | 94.1 | 4.7 |
| 1974 | 6,849 | 466 | 702.7 | 43.1 | 251.3 | 182.2 | 215.2 | 10.9 |
| 1975 | 7,195 | 602 | 984.7 | 61.8 | 366.6 | 162.9 | 373.6 | 19.8 |
| 1976 | 8,441 | 553 | 1,032.3 | 50.3 | 326.7 | 196.1 | 443.2 | 16.0 |
| 1977 | 8,173 | 757 | 1,498.2 | 78.7 | 473.0 | 286.0 | 636.8 | 23.7 |
| 1978 | 8,800 | 800 | 2,087.2 | 109.3 | 636.1 | 315.7 | 984.3 | 41.8 |
| 1979 | — | — | 2,652.2 | 134.8 | 864.3 | 340.8 | 1,247.6 | 64.7 |
| Percentage | | | | | | | | |
| 1970 | 100.0 | 2.2 | 100.0 | 9.3 | 60.2 | 15.6 | 14.1 | 0.8 |
| 1971 | 100.0 | 3.1 | 100.0 | 7.7 | 51.2 | 18.8 | 20.7 | 1.6 |
| 1972 | 100.0 | 4.1 | 100.0 | 7.9 | 41.1 | 16.2 | 33.8 | 1.0 |
| 1973 | 100.0 | 4.0 | 100.0 | 7.4 | 38.4 | 24.8 | 28.0 | 1.4 |
| 1974 | 100.0 | 6.8 | 100.0 | 6.1 | 35.8 | 25.9 | 30.6 | 1.6 |
| 1975 | 100.0 | 8.3 | 100.0 | 6.3 | 37.2 | 16.5 | 37.9 | 2.1 |
| 1976 | 100.0 | 6.5 | 100.0 | 4.9 | 31.6 | 19.0 | 42.9 | 1.6 |
| 1977 | 100.0 | 9.2 | 100.0 | 5.2 | 31.6 | 19.1 | 42.5 | 1.6 |
| 1978 | 100.0 | 9.0 | 100.0 | 5.2 | 30.5 | 15.1 | 47.2 | 2.0 |
| 1979 | — | — | 100.0 | 5.1 | 32.6 | 12.9 | 47.0 | 2.4 |

Source: Dahlman 1981.
[a] Excluding automobiles.
[b] Including automobiles.

scientific instruments (within transport, the largest item is motor vehicles). The subsectors which exported more than $10 million in 1970 and whose exports increased rapidly thereafter are special industrial machinery, office machinery, machinery not elsewhere classified, and motor vehicles. As can be seen, there was a significant change in the composition of capital-goods exports (broad definition) in the period.

## Exports and Government Policy: A Summary of Research

The export performance of the Brazilian capital-goods sector in the 1970s can be summarized as follows: a high annual rate of real export growth (28%— well above the country average of 11%) and a more than fourfold increase in the share of sector output exported. In an analysis of the role of learning and technological capacitation, it is important to assess to what extent this can be explained by direct export incentives and other measures having a direct effect on the volume and direction of trade.

One would expect a significant rise in the ratio of capital-goods exports to output to be associated with an increase in the profitability of exports relative to local market sales (given the overall profitability of the industry). Government policy would affect this ratio if export subsidies rose more than the level of protection. Furthermore, the policy variable which will have a direct effect on the growth of exports (and, via export demand, on the growth of output) is the effective real exchange rate (that is, the real exchange rate plus export subsidies). Two important studies, one focusing on capital goods and the other on exports in general, reached the conclusion that these incentives have not been sufficiently strong to explain the growth of capital-goods exports. Referring to the period 1969–1978, the first of these studies states that "while export incentives were important in these sales, their magnitude offset the overvalued cruzeiro. Moreover, there were no changes in the benefits provided during this period which can explain this growth."[8]

Tyler (1980) arrives at essentially the same conclusion for the 1970s. He also points out the existence of discrimination against exports which "can also be seen through a comparison of the protection afforded to production for the domestic market and the subsidization provided to export production" (Tyler 1980, 56–57). Both studies conclude that the development of Brazilian capital-goods exports was made possible by increased productivity or competitiveness, and they provide some direct evidence in support of this view (Tyler 1980, chap. 5; World Bank 1980, 39–43). An alternative hypothesis would be that in the 1970s export growth was motivated by an increase in the sector's excess capacity and that the price received probably covered marginal, rather than total, average costs. This hypothesis cannot be tested, but I suggest that its significance would be limited insofar as exports depend relatively less on price and more on quality, product specifications, and firm reputation, particularly, as will be seen later, for many capital goods. Finally, another factor which may have influenced exports is changes in demand. If anything, the local demand for capital goods increased during most of the period (a fact due to the investment demand of state-owned enterprises, see next subsection), and therefore its direct effect on exports would be negative.

What caused this increase in productivity and hence in exports? An answer to this question requires a close look at the supply and demand factors im-

---

8. World Bank 1980, pp. vi–vii. Nominal subsidies on most capital exports for 1975 ranged between 20% and 36% (see table 5.4); in 1978 the cruzeiro was regarded as overvalued by 25%–35%, and the extent of overvaluation had not changed substantially between the end of 1973 and mid-1978 (see World Bank 1980, 10, 11, and 13).

Table 5.3. Tariff and subsidy rates for capital goods (Brazilian Institute
of Geography and Statistics classification)

| | Average[a] tariff rate, 1976 | Export subsidies[b] | | |
|---|---|---|---|---|
| | | Fiscal subsidies | Tax rebates on capital goods | Total |
| Pumps and motors | 13.0 | 22.4 | 0.1 | 22.5 |
| Machine parts | 28.3 | 20.8 | — | 20.8 |
| Industrial equipment and machinery | 9.2 | 21.0 | 13.5 | 34.5 |
| Agricultural equipment and machinery | 11.7 | 16.8 | 10.9 | 27.8 |
| Office and domestic equipment and machinery | 17.1 | 24.9 | 2.0 | 26.9 |
| Tractors | 4.9 | 23.0 | 13.5 | 36.5 |
| Equipment for electric energy | 17.1 | 19.4 | 2.1 | 21.5 |
| Electric conductors | 20.5 | 20.0 | 1.4 | 21.4 |
| Electrical equipment | 34.4 | 20.2 | 4.5 | 24.7 |
| Electrical machinery | 28.1 | 25.7 | 4.6 | 30.3 |
| Electronic equipment | 18.8 | 20.0 | — | 20.0 |
| Communications equipment | 42.0 | 26.5 | 5.0 | 31.5 |
| Trucks and buses | 8.7 | 26.0 | 9.7 | 35.7 |
| Motors and vehicle parts | 26.5 | 25.8 | 4.3 | 30.1 |
| Shipbuilding | 0.1 | 23.5 | 5.9 | 29.4 |
| Railway and aircraft equipment | 2.1 | 15.5 | 7.3 | 22.8 |

Source: World Bank 1980, 74.
[a]Calculated as actual tariff collections divided by imports c.i.f.
[b]The fiscal subsidies via the IPI and ICM tax credits are estimates made with 1975 data.

pinging on the Brazilian capital-goods industry. The importance of skill acquisition and technological learning is suggested both by the changing array of commodities produced and exported by the sector and by the case studies of individual firms. Some of these studies show, first, that exports may be associated with increased capability of supplying more sophisticated products and, second, that this capability is in part a spin-off from earlier activities involving less sophisticated products supplied to the local market (see fourth section). It seems to me that these phenomena are important for explaining both the absolute and the *relative* growth of exports.[9]

9. Another factor that could explain the sector's export performance is a rise in the world price of capital goods supplied by Brazil; a basic issue here is the price index that should be used. The effects of a change in world prices cannot always be separated from the effects of technological learning. For example, a fall in the relative price of a simple good currently being exported may have a positive or a negative effect on exports according as the sector is or is not able to shift to a more complex good whose world price has increased.

Other measures favoring the capital-goods sector

The export performance of the capital-goods sector has been greatly influenced by government measures affecting the overall (private) profitability of capital-goods firms rather than by changes in the relative profitability of exports and local sales.

The measures may be divided into those increasing the domestic demand directed to local producers and those that favor the supply of capital goods (output and investment). The most important was the increase in government demand for locally produced capital goods. This followed from the expansion programs in the basic industries (e.g., steel, petroleum refining, petrochemicals) and infrastructure, in conjunction with clear directives to the state-controlled firms to purchase domestically produced equipment. I have found no figures on the extent of this demand nor on the proportion of total domestic capital-goods demand originating in government or government-owned enterprises. In some peak years before the present freeze in these programs, as much as two-thirds of the total demand for capital goods originated in government-owned enterprises. Other measures increasing demand include tariffs and non-tariff barriers to imports, tax exemptions, and cheap loans for the purchase of locally produced capital goods and other such incentives which are part of government-approved investment packages. Concerning tariffs, there are figures on the average tariff rate collected on capital goods for selected years between 1964 and 1977: they range between 6.9% and 16.2% (World Bank 1980, 16), the higher figure reflecting the increased protection of 1976 and

Table 5.4. Composition of investment-project approvals and share of local equipment required for approval: Brazil 1974 and 1978

| | Percentage of total fixed investment | | Required percentage of local equipment | |
|---|---|---|---|---|
| | 1974 | 1978 | 1974 | 1978 |
| Capital goods | 8.5 | 7.8 | 37.5 | 50.9 |
| Basic and intermediate metal industries | 23.1 | 61.0 | 48.4 | 75.9 |
| Chemicals and petrochemicals (including pharmaceuticals) | 20.9 | 11.4 | 48.7 | 97.6 |
| Nonmetallic, intermediate products, paper and cement | 23.0 | 11.0 | 54.3 | 80.5 |
| Automotive industry (including components) | 7.9 | 6.0 | 36.5 | 74.7 |
| Consumer goods | 16.5 | 2.8 | 30.5 | 56.5 |
| Total | 100.0 | 100.0 | 44.3 | 75.9 |

Source: World Bank 1980.

1977 that followed Brazil's balance-of-payments difficulties. Additional 1976 figures on tariffs can be found in table 5.3. Nontariff barriers to imports were much more important in increasing the demand for locally produced capital goods (the prior deposit required for importation was equivalent to 50% tariff). Table 5.4 shows the percentage of locally purchased capital goods required for government approval of special investment programs; as can be seen, this proportion has increased. Measures favoring locally produced capital goods include the investment incentives mentioned above (which favor several other sectors as well) and direct fiscal subsidies. I have found no figures for the rate of subsidy implied by the investment incentives, although some estimates may be available from Brazilian sources. Fiscal subsidies for the various products are shown in the second column of table 5.3.

### Some implications

The description presented above suggests that government has had a significant effect on the development of Brazilian capital-goods exports. Most of this effect, however, was indirect, that is, through a set of protection and subsidy measures whose immediate object was import substitution (or the avoidance of an unduly high rate of import growth) and the development of what economists would regard as an infant industry. These measures enabled the sector to accumulate physical capital, technology, and experience while for the most part supplying the local market.

Exports, whose growth accelerated significantly after 1973, should be regarded as a spin-off from what was essentially a domestically oriented activity (see next section). They should also be regarded as a consequence of underutilization of capacity, especially in the later years when, because of the energy crisis, the government cut back its expansion plans for basic industries.[10] In other words, two factors should explain the acceleration of exports: the capabilities accumulated during the import-substitution and infant-industry stages —physical capital, technological knowledge, experience, and skills—and the

10. Figures published in *Conjuntura* (July 1981) show that the rate of capacity utilization in the mechanical industry declined after 1976, ranging from 76% to 80% during 1977–1981 (these are simple averages of quarterly figures). The figures for 1971–1976 are closer to 90% and in 1973 the average stood above 92%. A broadly similar pattern is observed for the metallurgical and electrical, electronic, and telecommunication materials industry (the classification used here does not correspond to the standard industrial classification). Excess capacity in capital goods should be regarded as the "natural" outcome of a policy of promoting basic industries and infrastructure with an increasing reliance on local supplies. This assumes that once these programs are terminated, the shift to exports is not automatic and takes time. The excess capacity situation will be more serious if government cuts in investment are unexpected (as seems to have happened in Brazil), and if initial firm strategy was not explicitly geared to the eventual supply of exports.

subsequent reduction in domestic demand. This emphasizes the dual relationship between the local and export markets: in the short run they are substitutes; i.e., a reduction in domestic demand will, for a given level of capacity or output, tend to increase exports; on the other hand, stimulation of local demand (and/or protection of the local market) will, by making it possible to increase efficiency, eventually increase exports. In other words, the local market and exports may be regarded as complementary in the long run.

The microeconomic studies summarized in the next section suggest that an important technological learning or capacitation effect underlies the productivity increase in the Brazilian capital-goods sector. Together with enhanced firm reputation, this explains why a preliminary stage of supply to the local market has contributed to exports in the long run. It would be interesting to know the conditions under which an efficient export promotion strategy should conform to this pattern. Static trade theory suggests that if exports are to be encouraged at all, it should be via direct subsidization of exports (or devaluation). The Brazilian experience suggests that under certain conditions, export promotion need not be specifically geared to exports in the short run.

There would seem to be two such conditions. First, exporting is more difficult (or expensive) than supplying the local market. This may be due to the difficulty of achieving reasonable product quality, to the need to produce more complex or sophisticated products, to the greater firm reputation required, to the greater difficulty in understanding user needs, or to other factors, such as longer delivery times, supply of parts, or arranging for servicing. Second, the supply of goods to the local market has a spin-off benefiting exports, via the experience, capabilities, and reputation acquired.

These conditions will probably hold for some products and countries and not for others. For example, nonhomogeneous products such as machinery, instruments, and systems may satisfy the first condition since it may be inherently difficult for users to evaluate their functional utility or reliability. It will be even more difficult to export if the prospective purchaser runs the risk of substantial losses by making the wrong choice of product or supplier. In this case the local market will enable local firms to increase their world reputation as reliable suppliers of equipment, thereby eventually permitting them to penetrate export markets. A large domestic market is necessary, but not sufficient, for the second condition. The indirect strategy is probably less efficient for small or even medium-sized developing countries than for large ones such as Brazil, Mexico, and India.[11]

11. If both conditions hold, the efficiency of direct export subsidization of an infant industry or firm will be relatively low; i.e., both the absolute effect on exports and the ratio of exports generated to subsidies will be lower, at least relative to what it would be after the industry or firm has

## Technological Learning of Firms: A Summary of Some Interviews

The microeconomic interviews focused on learning processes and their relationship to export potential. This relatively narrow focus made it possible to interview a fairly large number of firms (given the time available) and to collect a large amount of information on the topic analyzed. Although a great deal of the information is qualitative (e.g., descriptions of endogenous firm spin-offs), some quantitative information was obtained on the economic value of the spin-offs. More important, a good deal of the groundwork for theorizing about the nature of the spin-offs and their economic value to the firm has been laid, and it has become much clearer what additional quantitative information will be needed.

### The firms interviewed

A total of eight firms were interviewed. In addition, background information on five other firms was obtained, and these firms can be interviewed in any follow-up of this project. Most of the firms belong to the capital-goods industry.[12]

A brief description of the product lines and areas of the firms interviewed follows:[13]

*Confab Industrial* was originally a boiler shop and has evolved into Brazil's most important welded-steel pipe producer. It also produces general industrial equipment (boilers, tanks, furnaces, etc.) and equipment for the steel* and cement industries.

*Dedini* is a group of companies having a dominant share of the market in equipment for the alcohol* and sugar* industries. It also produces equipment for cement* and steel* plants; boilers and turbines; steel rolls/bars for reinforced concrete, etc.

*Equipetrol* supplies pipes and accessories for petroleum exploration and production.

*Metal Leve* is a major supplier of pistons* and bearings for the automobile market. It also produces pistons* for diesel and aircraft engines.

*Promon Engenharia* is a very big engineering consultant offering com-

---

supplied the local market for a while. Note that indirect export strategy is not inconsistent with export or outward orientation even in the early phase. In addition, it does not imply that tariff or nontariff protection is preferable to output subsidization in that phase, nor does it exclude export requirements as part of the investment incentive package.

In the above discussion we reviewed the strategy for export promotion under the assumption that the country in question possesses a dynamic comparative advantage in the sector analyzed. It may, however, be impossible to separate the two issues.

12. The exception is Promon Engenheria, an engineering consultancy firm.

13. An asterisk indicates design capability or capability of offering complete installations.

prehensive engineering and management services for a variety of sectors (civil engineering,* construction,* petrochemical industry,* telecommunications,* transport, etc.).

*Treu-Máquinas e Equipamentos* is a relatively small company producing mixers* for the pharmaceutical, petrochemical, and nuclear-power industries and a wide variety of other special equipment for these and other industries (food, paint, etc.).

*Villares* is one of the most important locally owned industrial groups in Brazil. It produces automatic elevators,* motors, parts, special steels, steel-industry equipment, mining equipment, heavy capital goods,* transport equipment, overhead cranes,* and other goods. The group's original area was elevators.

*Zanini* is a group of companies originally focusing on sugar and equipment for sugar production. It later diversified into industrial equipment (cement plants, overhead cranes, etc.) and into equipment for the production of alcohol.*

The firms interviewed belong to a wide range of sizes [14] and cover a wide variety of capital goods and associated areas. For a number of reasons, however, the sample is not representative. First, there are some important capital-goods areas where no firm was interviewed. These include electrical equipment, automobiles or trucks, ships, machine tools, agricultural equipment, equipment for the pulp and paper industries, road building and heavy earth-moving equipment, and textile machinery.

Second, most of the firms either were leaders in one or more of the areas in which they were active (Confab in pipes, Dedini in sugar and alcohol-production equipment, Villares in overhead cranes and in a variety of other heavy capital goods, Promon in engineering services) or were important in these areas (Zanini in alcohol equipment). This reflects the view that relatively successful firms are those which have realized potential spin-offs from past activities and should therefore be interviewed when a qualitative understanding of the nature of such processes is aimed at. A follow-up sample would have to cover less successful firms as well.

Third, all the firms interviewed were locally owned. This is particularly problematic in view of the predominance of multinational enterprises (MNE) in some of the most important and fast-growing areas in the Brazilian capital-

14. Size measured by annual sales ranges from approximately $500 million (Villares) to $10 million (Treu). All firms interviewed stated that they currently have underutilized capacity (relative to "normal levels"). All the manufacturing firms which eventually exported had supplied the local market for some time before entering the export market.

goods industry (e.g., automobiles, trucks, and other equipment).[15] This is probably the main weakness of the present sample. The firms in the sample were chosen because they were successful or interesting (and available).

The main reason why no attempt was made to interview MNEs was my a priori belief that the main dynamic effect of MNEs is not technological learning but the externalities they generate in the process of supplier development (this would require a different focus). I have since revised my view about the learning processes in MNEs: although they have easier access to foreign technology (often from the parent company) than the locally owned firm, there need be no perceptible difference in the efficiency of learning and absorbing it.[16] Thus a study focusing on learning processes should also try to cover MNEs and consider external effects as well. Future research should therefore emphasize these firms at the expense of others.

### Endogenous firm spin-offs

The field work focused on identifying and conceptualizing endogenous firm spin-offs, principally technological capacitation or learning. Of all the indirect or dynamic effects of economic activity, it was felt that those materializing within the firm were the most direct and easily ascertainable.[17] These effects are also very significant and, given the imperfections of capital markets of less developed countries, may have important policy implications.

Spin-off from early products of the firm which significantly benefited subsequent products was detected in every one of the cases studied. In all but one case (where the only spin-off was enhanced firm reputation) significant technological learning was involved.[18] They also led to the emergence of export potential and to actual exports. It will be useful to summarize some examples that were reported (these are taken from the case studies completed for five firms of the sample).

The summary of each case starts by identifying the particular instance or

15. See Dahlman 1981, p. 12, table 4. However, Metal Leve, included in our sample, is an important producer of parts for the automobile industry.

16. I owe the distinction between access and efficiency to S. Teitel. He also pointed out that the pool of experience in their home countries enables the MNEs to develop suppliers in less developed countries more efficiently than locally owned firms.

17. A study illustrating this is Teubal 1982. This is not to deny that external effects are important. On the contrary, I am fully aware of their significance as well as of the difficulty of identifying them and of measuring their economic impact. A follow-up of case studies on MNEs would have to focus on some of the relevant externalities (see n. 16).

18. The exception was Equipetrol, where the founders had gained experience working in a similar firm. I shall try to identify the nature of this externality in a follow-up of the present study.

instances where learning or other endogenous spin-offs were identified. The mechanisms involved are then described: e.g., the manufacturing abilities learned in producing $X$ at $t$ make possible higher profits (than in the absence of $X$) from producing $Y$ at $t + 1$. An asterisk indicates that the spin-off has had an effect on exports. The remarks sections provide background and other relevant information (which may be supplemented by referring to the firm descriptions given above). It should be emphasized that the possibility of comparing different cases may be limited, owing to the heterogeneity of cases and to differences in terminology.

I.   Confab: equipment division
A.   *Learning or spin-off*
There is a natural progression in the design and manufacture of increasingly complex but related items of equipment such as varieties of tanks: floating roof tanks, simple pressure vessels, spheres, etc.
B.   *Mechanisms*
1. Similarity of equipment.
2. Accumulation of manufacturing capabilities: manufacturing methods, welding, metallurgical knowledge, etc.
3. Development of mechanical design capability, e.g., calculation of thickness relative to stress in tanks; selection of materials.
4. Development of a process design capability for certain types of equipment and industries (tanks, fire heaters, and cellulose processing plants).
C.   *Remarks*
Association with foreign companies is required; the typical sequence is, first, learning how to manufacture, then adapting to local conditions and, finally, the company adapting or improving without the licensor's help.

Each type of knowledge or capability represents a stage in mastering the technology. Mastery of the third stage (design capability), which may be very important for exports, requires the promise of a strong market.

II.  Confab: pipe division
A.   *Learning or spin-off*
Contribution of early types of pipe and operation of early plants to:
1. Planning and execution of new investment
(*a*) Layout.
(*b*) Speeding up plant installations (time was cut by half).
(*c*) Designing or specifying new machinery.
(*d*) Reducing errors in predicting the cost of investment programs.

2. Producing and selling pipes with more exacting specifications.

B. *Mechanisms*

1. Operating experience with early plants (A1, A2).
2. Experience in planning and executing investment programs (A1, A2).
3. Firm reputation effects critical for prequalification (A2).

C. *Remarks*

The firm installed 11 pipe plants, starting in 1961. The first two were designed and installed by the firm (including all the machinery) and produced water and petroleum pipes, respectively. The third plant produced pipes from steel coil (rather than steel plate) and was based on imported machinery. The modifications introduced in this plant made it possible to produce pipe casings for the expanding oil-well market and led to a completely new plant also based on steel coil. In the meantime, another huge plant using steel plate was set up, for which the company planned the layout and designed and constructed most of the machinery. This plant led the firm to develop special steels to reduce the thickness of pipe walls. Later plants produce pipes with more exacting specifications (e.g., coated for low temperatures, off-shore drilling, of X-60 or higher grade steel). This required new steels and new technology.

III. Dedini

A. *Learning or spin-off*

1. Diversification from the company's original lines (equipment for the production of sugar and alcohol) to new types of equipment: calcination plants (for the cement industry); basic oxygen furnaces and heat-recovery systems (steel industry); equipment for the paper and pulp, chemical and petrochemical industries.
2. Adaptation of boilers, turbines, and other equipment to other uses and markets, e.g., coal-fired boilers, turbines for the petrochemical industry.
3. In the original lines,* learning made possible maximum exploitation of market growth after 1973, both at home and in other Latin American countries.

B. *Mechanisms*

1. Accumulation of manufacturing capabilities (A1, A2, A3).
2. Similarity of equipment: heat-recovery systems for steel plants are similar to boilers; speed reducers for calcination ovens (cement industry) are similar to those used in sugar plants (A1, A2, A3).
3. Development of design capability:* especially important for sales to nonsophisticated users both at home and abroad (A3).

C. *Remarks*

The motive for diversification was market trends. The decision to produce steel equipment was based on manufacturing capabilities (e.g., the firm decided not to produce blast furnaces). In both steel and cement equipment, the company acquired a design capability that enabled it to offer complete plants (in association with a Japanese firm) (A1).

There was no special difficulty in adapting boilers and turbines associated with sugar and alcohol. Coal-fired boilers, in which the company is quite successful, use foreign technology. The company has also made know-how agreements in the area of turbines (A).

IV. Metal Leve

The firm was established in 1950. Its first activity was import substitution (automobile pistons). Today it is a major producer of pistons for motor vehicles, aircraft pistons, piston pins, plain bearings, and bushings.

A. *Learning or spin-off*

   1. Domestic supply of aircraft pistons made exports possible.*

   2. Exports of aircraft pistons raised overall productivity.

B. *Mechanisms*

   1. Access to foreign know-how:* the company's success in developing and supplying aircraft pistons to the local market enabled it to obtain a license from a foreign company, which helped in its initial export efforts. Foreign firms were unwilling to provide the technology before the firm succeeded in the local market.

   2. The introduction of quality control: obtaining FAC approval for sales in the United States involved the introduction of formal quality-control procedures, which made it possible to mass-produce high-quality bearings and pistons

   3. Emergence of a design capability: this probably benefited from quality control. The recently established R & D center was originally staffed largely by personnel from quality control (A2).

   4. Firm reputation (A1).

C. *Remarks*

The value of the spin-off is high since there is a trend to stricter specifications in response to the energy crisis, even though significant complementary investments in physical capital are required.

The technological base, including the capability acquired, enabled the firm to participate in a joint development effort in pistons for alcohol-burning engines. (Such pistons require tougher materials and a different design, but are not exceptionally difficult to produce.)

An important factor in the firm's development was its forecast that market trends would favor heavy-duty pistons, which are more difficult to produce (they are larger and require iron-nickel insert and more precise work because the engines work closer to their theoretical limits).

V.  Promon Engenharia
This engineering consultancy firm was founded in 1960. Its initial area of activity was in the process industries (e.g., petroleum refining). Since then it has diversified into aluminum, hydroelectric power, transport, and telecommunications. In 1975 the firm started exporting its services and set up an independent R & D facility.
A.  *Learning or spin-off*
    Learning occurred:
    1.  Within a particular area, such as the civil engineering aspects of hydroelectric power, where increasingly complex projects were undertaken.
    2.  Across different areas: early activity in process-industry projects helped in other engineering projects such as dams.
B.  *Mechanisms*
    1.  Accumulation of a design capability (A1, A2).
    2.  Development of a capability for total project management* (A2).
    3.  Firm reputation.*
C.  *Remarks*
    The firm always entered a new area in association with a foreign partner, from which it acquired a design capability that helped it in subsequent projects (A2); in particular, its ability to offer turnkey plants enabled it to export (B2). Exports also benefited from the firm-reputation effect (B3). The firm started exporting hydroelectric projects in 1981 (to Chile, Costa Rica, and Nigeria).

VI.  Treu
The firm started in 1943, supplying the pharmaceutical industry; in 1968–1969 it began to supply the emerging petroleum and petrochemical markets, and since 1979, the nuclear-power industry.
A.  *Learning or spin-off*
    1.  Supply of mixers and other equipment to the petrochemical and nuclear-power industries benefited from the firm's prior experience in supplying the pharmaceutical industry with similar products.
    2.  Supply of equipment of various types to the petrochemical industry helped the firm to sell equipment to other industries (e.g., mixers

to the paper and cellulose industry; small mills to the ink, cocoa, and food industries).

B. *Mechanisms*

    1. Manufacturing experience (A1).

    2. Similarity of equipment: the only difference between different mixers was their size (3 HP for the pharmaceutical industry; 3–300 HP for petrochemicals; 3000 HP for the nuclear-power industry). Methods of manufacture and quality control were similar, for example, in machining, casting, and welding. Experience with smaller equipment reduces the risk in manufacturing bigger equipment (A1).

    3. Firm reputation was critical for entering the nuclear-components market (A1).

    4. Accumulation of a design capability: this evolved at a later stage and helped in the adaptation of equipment to other industries (A2).

C. *Remarks*

The first trained engineers to be hired (i.e., other than the owners) were engaged in 1970, and quality control and process engineering were developed (B4). The firm predicted that the new nuclear market was coming and prepared itself by investing in a large new tank for full-size mixer testing. The new market reduced excess capacity (i.e., the economic value of the spin-off from past activity is likely to be high) (A1).

Summary of results

Eleven endogenous spin-off mechanisms—most of them related to learning in one way or another—were detected in the firms interviewed. It may be useful to group them into those related to manufacturing, those related to design, and others.

*Learning related to manufacturing:* (*a*) the accumulation of manufacturing capabilities (in most firms); (*b*) operating experience (in Confab's pipe plants); (*c*) experience derived from similarity of equipment (e.g., Treu's mixers of increasing size); and (*d*) improvement of quality-control procedures (e.g., in Metal Leve for aircraft pistons).

*Learning related to design:* (*a*) development of a mechanical design capability (e.g., Confab's equipment division); (*b*) development of a process design capability;[19] (*c*) development of a total project management capa-

---

19. It may sometimes be difficult to separate the two types of design capability; in other cases, the distinction is less relevant.

bility[20] (e.g., Promon's hydroelectric power projects and Dedini's alcohol plants).

*Other mechanisms*: (*a*) enhanced firm reputation (e.g., Metal Leve, due to its success in supplying aircraft pistons to the local market); (*b*) improved capability to plan and execute investments (Confab, pipe division); (*c*) increased knowledge of markets.[21]

The most common spin-off mechanisms are related to manufacturing capabilities: they were present in all firms except consultants such as Promon; they seem to have been extremely significant in some cases; and they generally emerge before design capabilities. Operating experience, which was relevant to Confab's pipe plants, refers to the operation of whole plants (e.g., aspects of layout) and should therefore be distinguished from the accumulation of manufacturing capabilities on the part of individual operators. This factor is presumably more important in continuous or semicontinuous production processes than in assembly-type industries (more knowledge about the differences in technology between the two types of industry is required before we can be sure about this). An important implication is that operating experience directly improved the capability to plan and execute investments; that is, the efficiency of investment is not only a function of past investments, a feature that has been observed in microstudies of steel plants.[22] The phenomenon of capacity stretching is also to a large extent a result of operating experience.[23] A final point is that in manufacturing, the capability to produce products with more exacting specifications (or heavier items of equipment) followed from the production of simpler and smaller items.

A natural sequence of events in successful capital-goods firms seems to be, first, the acquisition of manufacturing capabilities and, second, the acquisition

20. Including capability to supply turnkey plants. This includes, but goes beyond, the capabilities associated with mechanical or process design.

21. The spin-off processes indicated here differ from those reported in Teubal (1982) for a successful electronics firm. These were recorded at two levels: within generations of a given product class, and between one product class and another. The phenomenon of equipment generations does not occur in many of the mechanical goods produced in the Brazilian capital-goods sector. (The rapid substitution of one product class for another also occurs rarely.) Moreover, in electronics, the most significant spin-offs were related to market feedback and R & D, rather than to manufacturing capabilities. These differences reflect other more basic differences between the highly dynamic electronics systems/equipment areas and the metalworking industries.

22. See Maxwell 1976 and Dahlman and Valadares Fonseca 1978. Both studies were conducted within the ECLA/IDB/IDRC/UNDP Research Programme on Scientific and Technological Development in Latin America.

23. For capacity stretching see Hollander 1965; Katz et al. 1977; Maxwell 1976; Dahlman and Valadares Fonseca 1978.

of design capabilities in a narrower set of products. Sometimes, although this should be further documented, the link between the two is the procedures, equipment, and personnel involved in quality control (this is what seems to have happened in Metal Leve, where quality-control personnel staffed the R & D laboratory set up by the firm; it seems to have happened in the Usiminas steel plant too). The interviews also suggested that the full development of a design capability may be jeopardized by a reduction in the local demand for the product, apparently because of the fixed costs incurred.[24]

A very important distinction suggested by some of the examples is that between *simple or process learning* and what might be called *product learning*.[25] The former enables the firm to produce existing equipment at lower cost or to introduce relatively minor technological improvements. The latter, on the other hand, enables it to shift to new, but in some respects technologically related, types of equipment which are more difficult to produce or which require a design capability as well. Simple learning was accumulated by Dedini from its original lines of sugar-processing and alcohol-production equipment. Simple learning enables the firm to exploit fully the (temporary) expansion of its existing market. Most of the other instances of spin-off are due to qualitative learning. They involve either more exacting specifications (e.g., Confab's X-70 grade pipes, Metal Leve's diesel-engine pistons) or the manufacture of larger items (Treu's mixers) or a design capability, or a capability for offering complete packages (Dedini's turnkey alcohol plants, Metal Leve's diesel-engine pistons, or Promon's project management and turnkey projects).[26]

The essential difference between the two types of learning is that while simple learning makes possible fuller exploitation of the market(s) and user segment(s) currently served by the firm, product learning enables the firm to better exploit the existing set of submarkets or user segments related to its area of expertise. Thus, it does not have to be dependent on expansion of the specific submarket/user segment served in order to increase sales. Moreover, it will be better able to exploit new opportunities in the world and domestic markets and to avoid some of the pitfalls. The policy implications of this distinction for exports and for industrial development may be far-reaching.

It is interesting to note that in all cases except one (Treu) spin-offs involving product learning have led to exports. In some cases, the world market requires

24. A consequence of this may be reduced export capability at a subsequent stage.

25. Thanks to Larry Westphal for having suggested the terms used.

26. While our simple learning category is part of Lall's (1980) elementary learning category, there is no similar correspondence between our product learning and his intermediate and/or advanced learning categories. Thus product learning includes transitions to the manufacture of more complex or sophisticated products. This may occur *without* process or product adaptations (from a world point of view) and therefore would be considered by Lall to be elementary learning.

more sophisticated products than those currently supplied to the domestic market (some of Confab's pipes made of high-grade steel and piping for transporting fluorine, for which the firm has obtained prequalifcation [see next paragraph], and possibly some of Dedini's turnkey alcohol plants or some of Promon's turnkey projects).[27] In other instances the firm is exporting less sophisticated products than those it currently supplies domestically (Promon's exports of hydroelectric projects), but even the less sophisticated products were the result of product learning while serving the domestic market. In some cases, it seems that the reputation acquired by supplying sophisticated products to the local market (e.g., nuclear-power components, whose production was made possible by qualitative learning) opened up some export markets for less sophisticated products.[28]

No instance was detected of exports which were not preceded by sales of the same or similar equipment to the local market. The acquisition of firm reputation was a major reason why supply to the domestic market was important for exports. Enhanced reputation enables the firm to *prequalify* for the supply of goods in international tenders (Confab, Promon) or to gain access to a foreign license which would enhance its export acceptance (Metal Leve). Another important spin-off from supplying the local market was learning how to assemble and offer turnkey projects or sophisticated products, a critical factor in some (especially less sophisticated) export markets (Promon).[29]

Table 5.5 presents some qualitative information pertaining to the economic value of the spin-offs reported earlier. Very little quantitative information is available at this stage, so the table focuses on the factors mainly responsible for a "high" or "low" economic value (as reported in the interviews). The preponderance of cases with a high economic value reflects the fact that the sample includes relatively successful firms. Needless to say, there are some serious conceptual problems in determining the value of the spin-off flowing from a particular activity.[30]

27. This includes situations where the trend in (rather than the state of) international markets favors more sophisticated products.

28. It is not clear whether the examples given here contradict the Linder hypothesis. In both cases, the domestic market is a proving ground for export production, primarily because of skill acquisition (and firm-reputation effects), and not necessarily similarity in demand.

29. Although there are instances in other countries of capital-goods exports without a prior phase of supply to the local market (e.g., ships from Korea), my hypothesis is that it is much more difficult with capital goods than with manufacturing in general.

30. The value of the spin-off from product *i* to product *j* is the difference between profits actually obtained from *j* and those that would have been obtained in the absence of product *i*. The latter is problematic because in the absence of *i* the firm might not undertake *j*. The formula for calculating the value of spin-offs will differ in each case.

Table 5.5. The economic value of spin-offs

| Spin-off and main contributory factors | Value of spin-off |
| --- | --- |
| **Confab, equipment division**<br>Reduced local market for equipment before full mastery of a design capability. | Not high |
| **Confab, pipe division**<br>Market trends favored a more sophisticated pipe. There was a very significant reduction in risk and in time for planning and executing investments (a newcomer needs eight years for a new pipe plant; Confab may need half that time). A newcomer incurs high costs of purchasing know-how. | High |
| **Dedini**<br>*Diversification from original activities* (A1):[a]<br>Narrow market, at least up to now. Possibility of losing part of design capability (no economic justification for maintaining a large team of engineers). | Low |
| *Diversification within traditional lines* (A3):<br>Shorter delivery times (1½ years less than new entrants) in a market predicted to reach satiation in a few years. Although the technology is not very complex, purchase of know-how is expensive for new entrants—with Dedini's volume, savings could reach $22 million. | High |
| **Metal Leve**<br>*Spin-offs from aircraft pistons* (A1, A2):<br>Market trends favored high-quality pistons with more exacting specifications (e.g., heavy-duty diesel-engine pistons). | High |
| **Promon Engenharia**<br>*Increasing complexity of projects undertaken* (A1):<br>It is too risky to undertake such projects without prior experience with simpler projects. | High |
| **Treu**<br>*Spin-off from mixers* (A1):<br>It is too risky to produce big mixers without previous experience with smaller ones. The market for big mixers expanded and that for smaller ones contracted—the firm would have had much more excess capacity if it had not entered the new market. The market for bigger mixers is less competitive than the market for smaller ones. | High |

[a]The numbering refers to the case histories given above.

Except in cases of simple learning and in firms providing services, the materialization of potential spin-offs from a particular activity requires complementary effort and investment (the exceptions were Dedini and Promon). Some of these efforts and investments should be planned and executed in advance, and this may require successful market forecasting on the part of the firm. For example, in the 1960s Metal Leve complied with Caterpillar's H-D piston requirements. The firm was particularly successful in the 1970s, when market trends favored the H-D over the gasoline-engine pistons, which had accounted for a majority of sales a decade earlier. Others require heavy investment in physical capital and involve considerable risks (Confab pipes; Metal Leve). Finally, know-how agreements with foreign firms may be a necessary condition for shifting to a more complex, but technologically related, product (Treu). This situation also requires the introduction of more stringent quality-control procedures and investments in instrumentation for quality control (and/or testing facilities).[31]

## Conclusions

The Brazilian capital-goods industry experienced a very high rate of export growth during the 1970s together with a significant rise in the share of output exported. The sector's output and exports both grew faster than the output and exports of the manufacturing sector as a whole. A brief survey of existing work on government policy towards the sector shows that export subsidies did not unambiguously increase during the period, that they only compensated for an overvalued cruzeiro, and that they did not discriminate in favor of the export market. Therefore changes in these instruments of government policy do not provide an adequate explanation of the sector's export performance, and other explanations must be sought. Previous work has assembled quantitative information on the costs of production of various capital-goods product classes. This paper has provided information on endogenous firm spin-offs in a sample of capital-goods firms. The most commonly found and significant spin-offs involve the accumulation of manufacturing capabilities (learning to manufacture), with the development of design capabilities following in certain cases. A nontechnological spin-off was enhanced firm reputation; all the firms interviewed believed that this played an important role in their growth. The interviews also suggested that exports benefited from the increased capability to supply either more sophisticated products (products with more exacting specifications) or turnkey plants. There were, in addition, important reasons why,

---

31. A more detailed description of complementary efforts can be found in the complete write-ups of the firm interviews. These are available from the author on request, subject to permission from the firms.

in the firms analyzed, exports of capital goods benefited from prior activities associated with supplying the local market. It would be interesting to know whether capital-goods exports of other countries, such as Korea, developed *after* learning and enhanced reputation were acquired while supplying local markets.

The pattern indicated here may not be applicable to the capital-goods sector as a whole, nor does it prove that capital-goods exports must, on efficiency grounds, follow a stage of supplying the domestic market. It only suggests the need to link exports (and the emergence of export potential) with the theory of the infant firm.[32] The resulting framework would deal with the following issues.

*Strategy for promoting exports*: The characteristics of most capital goods suggest that the export-promotion strategy for those goods whose production is justified in the long run should be indirect, at any rate in large LDCs. In phase 1, prior to the acquisition of manufacturing and design capabilities and firm reputation, infant-industry measures (protection, subsidization, or whatever meets the case) should be implemented. Direct export subsidization will fail to promote exports or at best will do so only at very high cost, especially when exports require greater technological sophistication than is required to supply a less-demanding local market and when export firms have to be prequalified (as is often the case with capital goods, especially heavy capital goods). The effectiveness of export promotion will increase in phase 2 once the firm has acquired sufficient experience and reputation. This strategy would seem to be applicable to other experience goods, where the economic consequences of inappropriate choice are likely to be high, and in situations where (*a*) exporting is significantly more difficult than supplying the domestic market, and (*b*) the latter generates a spin-off benefiting the former.

*Infant industries—price versus other incentives*: The preference for output subsidization over other incentives, derived from static trade theory, is far less unambiguous when firm-reputation effects are also critical in phase 1. Under these conditions, even very high output subsidies may do very little to stimulate output (and, indirectly, learning). It follows that, given that competition basically rests on the user's confidence in the supplier, which the local firm does not enjoy in phase 1, there may be a case for nontariff protection (such as the Brazilian domestic-input schedules imposed on firms receiving government support). Considerable effort is still required to determine the conditions under which it would be optimal. Factors such as market size and the efficiency and learning motivation of entrepreneurs may be critical. Any optimum solution should take these aspects into account as well.

Finally, the learning and spin-off processes observed do not seem to be

---

32. Thus confirming the overall approach followed by Katz and Ablin (1976).

those which are traditionally used in productivity studies. For example, use of the weighted average of a firm's past R & D efforts as an indicator of its stock of knowledge[33] does not seem to be adequate for capital-goods firms in developing countries, since formal R & D activities, if undertaken at all, are rarely initiated by a newly established firm.[34] Significant learning has nevertheless taken place, and some of it (product learning) is best represented by cumulated output. Skill acquisition in both manufacturing and design seems, at least in some cases, to be related to the firm's current degree of sophistication (rather than to cumulated output) and to current investment in upgrading and purchasing or absorbing new technology.[35]

## References

Dahlman, Carl J. 1981. Untitled study on Brazilian capital goods. Washington, D.C.: World Bank. Dahlman, Carl J., and Fernando Valadares Fonseca. 1978. "From Technological Dependence to Technological Development: The Case of the Usiminas Steel Plant in Brazil." ECLA/IDB/IDRC/UNDP Research Programme on Scientific and Technological Development in Latin America (hereafter abbreviated Res. Prog. Sci. Tech. Dev. Latin America), Working Paper no. 21. Buenos Aires.

Griliches, Zvi. 1973. "Research Expenditures and Growth Accounting." In *Science, Technology, and Economic Growth*, edited by B. R. Williams, 59–83. New York and Toronto: Wiley, Halstead Press.

Hollander, S. 1965. *The Sources of Efficiency Growth: A Case Study of the Du Pont Rayon Plants.* Cambridge: MIT Press.

Katz, J. 1980. "Domestic Technology Generation in LDC's: A Review of Research Findings." Res. Prog. Sci. Tech. Dev. Latin America; Working Paper no. 35. Buenos Aires.

Katz, J., and E. Ablin. 1976. "Tecnologia y exportaciones industriales: Un analysis microeconomico de la experiencia Argentina reciente." Res. Prog. Sci. Tech. Dev. Latin America, Working Paper no. 2. Buenos Aires.

33. See Griliches (1973) for the use of this variable at the aggregate level. Subsequent studies by him and others have used it at the sector and firm levels.

34. We have already noted that the acquisition of design capability does not imply R & D. In any case, there are usually no consistent data on R & D expenditures. This study further reinforces an important implication of microeconomic case studies of firms, namely, that technological effort should not be identified with R & D (see Hollander (1965); and the studies conducted under the ECLA/IDB/IDRC/UNDP Research Programme on Scientific and Technological Development in Latin America; also Katz 1980).

35. The case of Treu suggests this in connection with the size of capital goods. The larger the product currently supplied, the easier it is to acquire the technology and additional skills needed to supply other large or larger items. This pattern can be found in other firms of the sample with respect to other product dimensions (e.g., precision requirements in pipes and pistons; complexity of engineering projects).

Katz, J., M. Gutkowski, M. Rodrigues, and C. Goity. 1977 "Productividad, tecnologia y esfuerzos locales de investigacion y desarrollo." Res. Prog. Sci. Tech. Dev. Latin America, Working Paper no. 13. Buenos Aires.

Lall, S. 1980. "Developing Countries as Exporters of Industrial Technology." *Research Policy* 9:24–52.

Maxwell, Philip. 1976. "Learning and Technical Change in the Steel Plant of Acindar S.A. in Rosario, Argentina." Res. Prog. Sci. Tech. Dev. Latin America, Working Paper no. 4. Buenos Aires.

Rosenberg, N. 1976. *Perspectives on Technology.* Cambridge: Cambridge University Press.

Teubal, Morris. 1982. "The R & D Performance through Time of Young, High-Technology Firms: Methodology and an Illustration." *Research Policy* 11:333–46. (Chap. 6 in this volume.)

Tyler, J. 1980. *Advanced Developing Countries as Export Competitors in Third World Markets: The Brazilian Experience.* Washington, D.C.: Center for Strategy and Advanced Studies, Georgetown University.

World Bank. 1980. "Brazil: Protection and Competitiveness of the Capital Goods Producing Industries." Report no. 2488-BR. Washington, D.C.

World Bank. 1981. "Country Economic Memorandum: Brazil." Report no. 3275a-BR. Washington, D.C.

This study was made possible by a grant from the Inter-American Development Bank. I am extremely grateful to Hugh Schwartz for his comments, help, and suggestions throughout its execution. I appreciate the help of Carl Dahlman in providing both important material and useful comments on a wide variety of issues addressed by the study. The collaboration of Sylvio de Aguiar Pupo, Helio Nogueira de la Cruz, Nuno de Figueiredo, and Jorge Katz is much appreciated. Nadav Halevi, Ruth Klinov-Malul, Francisco Sercovitch, and Simon Teitel made useful suggestions on various points. I am deeply grateful to the following people for allowing me to interview them: J. L. de Almeida Bello, Marcelo Ferraz de Amaral, Paulo do Amaral Gurgel, Adolfo Anunziata, Ayrton Bassani, German Bensadon, Alberto Pereira de Castro, Tobias Cepelowicz, Nicolino de Cillo F°., Ronald Jean Degen, Amantino Ramos de Freitas, Israel Gochnarg, Edgardo Hernandez, Sergio E. Mindlin, Agustin J. Pazos, José Stamile Piquet Carneiro, Eustaquio Reis, Luis Rico, Francisco Matias Silvano, Joaquim E. Cirne de Toledo, Sergio C. Trindade, J. Tyler, and Carlos R. Villares. Lastly, I am grateful to Susanne Freund, who provided very useful editorial assistance. The field work for this study was conducted in July–August 1981.

# 6 The R & D Performance through Time of Young, High-Technology Firms
## Methodology and an Illustration

## Morris Teubal

### Introduction

This paper summarizes the pilot stage of a microeconomic study of the R & D performance through time of young, high-technology firms in the Israeli electronics sector. The objects of the study are (1) to develp a methodology or framework for organizing the information on R & D input and output from the early history of each firm with respect to a set of R & D projects (innovations) in a way that may be useful for subsequent analysis; (2) to build a data base of R & D projects (and firms) in accordance with the procedures developed; and (3) to analyze the data. The last would in turn involve estimating the direct rate of return of each project (or of the R & D of each project) and its variation over time within each firm and identifying the factors explaining commercial performance, bearing in mind two sets of variables: area variables, such as market size, which are exogenous to the firm, and firm-related variables, especially those associated with learning and the acquisition of skill (e.g., interproject synergy).

The project will further the gradual building of a data base of innovations and should set the stage for an analysis of success and failure in innovation. The present work differs from existing data bases and analyses in three respects: (1) both commercially successful and failed innovations are included, (2) a continuous measure of the direct[1] rate of return to R & D (or project R & D productivity) is used instead of the success-failure dichotomy, and (3) chronological time is explicitly considered.

Most studies on innovation performance at the project level have either considered successful innovations (Myers and Marquis 1969; Utterback 1975) or divided the sample into successful and failed projects (Science Policy Research Unit 1972; Rothwell et al. 1974; Teubal, Arnon, and Trachtenberg 1976). In

---

1. This measure excludes the indirect effect of a project on the profitability of other projects.

this study, a quantitative (and continuous) measure of performance is used instead.[2] Probably the most significant departure of this project from previous studies of innovation performance is that we explicitly consider chronological time. We are interested in comparing the rate of return to R & D in project 2 executed at time $t_2$ with that of project 1 executed at time $t_1 < t_2$. We are also interested in the indirect contribution of project 1 to project 2, such as when they are successive generations of a given product. This set of issues links the innovation-performance literature with the infant-industry literature.

Although information was assembled on three firms at the pilot stage of the research, this paper can report only some general results for one of them —a successful firm designing, producing, and selling electronic instruments and systems. This will illustrate the general approach of the paper and the methodology developed so far.[3] The next section describes the process of assembling and organizing the information on projects and firms. It also describes some constraints in revealing this information. Next, I discuss some methodological issues, in particular the definition of *R & D project*; the following section illustrates the approach and presents estimates of project rates of return and a preliminary analysis. The paper closes with some general remarks on the merits of the approach presented here.

### Collecting and Organizing R & D Project Data

The collection and organization of data on individual R & D projects required for analytical purposes are subject to a number of difficulties which need separate consideration. Those encountered in our project include the fact that the information required is not always available in a suitable form and that in any case firms are often reluctant to provide information. Also, a working definition of *R & D project*, or *innovation*, had to be devised. We had access to files containing general information on the products being developed by various firms as well as consistent annual series of total R & D inputs into the development of the products listed. The products in any one file do not necessarily belong to the same R & D project; if they do not, the project breakdown cannot be obtained from the file. We also had access to some information on sales, which suffers from the same shortcomings as the R & D data in the files; and there is no systematic information on production and marketing costs.

The recorded R & D input and sales data were supplemented by inter-

2. Mansfield et al. (1977) have also reported rates of return to 17 innovations of "average or routine importance," both direct private and social, in order to ascertain the extent of the gap between them. We shall be reporting the direct and indirect component of the private return (see tables 6.1 and 6.2 and below) with a view to explaining commercial performance *through time*.

3. Research on a second firm was discontinued at a later stage because of lack of cooperation from management. Research on a third firm proceeds but at a slow pace.

viewing the firms concerned, but we have not at this stage asked for data on production and marketing costs. As mentioned, there was some reluctance to respond, in part because the information asked for was not always available in a suitable form. A second point in this connection is that on the whole, the more successful firms were better disposed than the less successful. These interviews of course also produced much of our qualitative information, which firms were quite willing to provide.[4]

The first object of collecting qualitative information was, as mentioned earlier, to group products into projects in such a way that we could feel confident that we were allocating the R & D input and output data to entities which we understood and which also had some relation to the objectives of the analysis. It also in principle makes it possible to regroup products into a new set of R & D projects in response to a change in the objectives of analysis (see below). As a result of these considerations an attempt was made to organize the qualitative story of individual products (or groups or projects) around a common set of issues.

*The motivation for undertaking the project.* Under this heading we distinguish between "need pull" and "supply push." The former involves a description of the state of the art and in particular of needs not completely satisfied by current technology, the competitive pressures facing the firm, the innovation opportunities opened up by new technology, and the firm's forecasts of market size and sales. In both, an attempt is made to determine the new opportunities opened to the firm as a result of its activities with earlier products (synergy, skill acquisition, and firm-reputation effects).

*The innovation process.* Here we briefly summarize the main highlights— alternative design approaches and the one eventually selected, stages completed (with dates of completion), special difficulties encountered, the purchase or other transfer of knowledge, and the testing of prototypes, including feedback from users.

*The products (or systems) developed.* We determine their uses and their users, and sometimes the firm's advantage (or disadvantage) as compared with competitors or with the firm's previous products.

*Area variables.* Differences in area variables presumably play a role in explaining differences in the commercial performance of R & D projects. At this pilot stage of the study we have collected whatever recorded information was

---

4. It should be emphasized that these data problems are specific to studies whose unit of analysis is the individual innovation or R & D project. Most of the economics literature on R & D (particularly in the area of industrial organization) is conducted at the level of the industry or the firm and not at the project level. Such data problems have thus not arisen, and others, such as the definition of R & D, have been prominent. The previous literature analyzing R & D projects has too often avoided discussing the data and conceptual problems raised here.

available and what was obtainable from interviews. However, the information is by no means complete and will have to be supplemented at the postpilot stage, for which an analysis of commercial success and failure is planned.

### Constraints on publication

The qualitative and quantitative information collected for each firm was organized into a separate file. Each file includes a section on the overall development of the firm, a section on its R & D projects, and a section on firm strategy and area variables. The file for the electronics firm reported here includes approximately 220 typed pages summarizing the information collected during a period of seven years. Prior to organizing the file in 1980, we had prepared two case studies of the firm, one at its fourth year and the other at its eighth year of existence. These were never published or widely circulated, but the information was later incorporated into our file.

During the above-mentioned period, several interviews with management were conducted, and in addition we had access and made use of written information on the products/projects of the firm. (The structure of the interviews followed the structure of the firm file described above, but each interview covered a different set of projects.) In addition to interviewing management, we interviewed three important local users of the products included in this paper. This enabled us to check some of the qualitative information provided by management, especially on the functional performances of the various products, on their degree of acceptance in the local market, and on the degree of acceptance they had or were likely to have in foreign markets.[5] Complementary qualitative information on the various products/projects was obtained from technical people who, although involved in some way with a particular project, were not part of the company's permanent staff. As a result of the information gathered from other sources, we confirmed our feeling that the opinions expressed by management were reasonably close to reality.

After completing the file for the firm considered, a first draft of the paper was written in 1981. It included *all* the relevant qualitative and quantitative information on the subset of products/projects included. The presumption was that the firm would have no objection to publication since a point was made of *excluding* the most recent projects of the firm. Despite this, and contrary to our expectation, authorization for publication was not granted. The information presented in the remaining part of this paper is the result of a second draft that removed all reference to the firm, its product lines, dates, and project information (both quantitative and qualitative).

---

5. We had from the very beginning been impressed by the disposition of management to admit past errors and failures, as well as to present the current difficulties and risks confronting the firm.

## The Grouping of Products into R & D Projects

The set of projects of a firm should be that grouping of products[6] which is optimal in the light of the objectives of analysis and the limitations of the data. Since we have already referred to the latter, we concentrate here on the former.

The collection and organization of firms' R & D data and their analysis have as their first object the study of the R & D performance through time of young, high-technology firms. This involves, first, describing the changes in direct R & D performance (e.g., whether it is increasing or decreasing) and, second, explaining the pattern obtained. The explanation should focus, at least initially, on endogenous firm variables such as synergy in R & D and marketing and learning by doing. The reason for this is not only that information on these factors is probably more readily available to the analyst than data on the details of the export markets (area variables) which the firms serve, but also because the character and strength of these processes are a major, if not the major, determinant of performance.[7] It follows that we will attempt to focus on the *indirect effects* of the products launched (and activities undertaken) in a particular year on the profitability of products launched subsequently.

This is a convenient point for a digression on the concept of synergy. In an extensive discussion, Ansoff (1968) defines it as the "joint effects . . . resulting from the addition of new product-markets to the firm" (p. 28), or as the "$2 + 2 = 5$" effect (p. 75). The present paper refers to specific products or to R & D projects (say $x$ and $y$) and distinguishes the contribution of $x$ (inputs, outputs, or both) to the profitability of $y$ from the contribution of $y$ to the profitability of $x$; that is, joint effects are not dealt with. When project $x$ is referred to time $t_1$ and project $y$ to time $t_2 > t_1$, the emphasis is on the contribution of $x$ to $y$ and not the other way round (we provide some quantitative estimates of this effect). In addition to synergy between products belonging to a given project (intraproject synergy) we refer to the synergy between the products of one project and those of another (interproject synergy).

Synergy in R & D and in marketing are both considered. R & D synergy between product $x$ (developed during period $t_1$) and $y$ (developed at $t_2 > t_1$) occurs when the R & D associated with $y$ is cheaper or more efficient (or both) as a result of prior R & D associated with $x$. Marketing synergy may be due to a variety of factors: lower fixed start-up costs (since some were already in-

---

6. It is not relevant to our present concern to go into the distinction between products, systems, components, and so on, and we use the term *product* as the general term throughout.

7. This is not to minimize the effect of a changing environment on the pattern of R & D performance. However, in our view, how effectively a firm can respond to such changes depends on the know-how base, the experience and skills accumulated up to the relevant point in time. See the following section.

curred in connection with $x$); lower variable marketing costs (as when the existing sales force selling $x$ adds $y$ to its line); greater sales of $y$ due to user satisfaction with $x$ (a firm-reputation effect) or to the fact that the demand for $y$ is greater when offered in conjunction with $x$ than in its absence (line effect); and so forth. Marketing synergy is termed sales synergy by Ansoff (1968, 75), while our R & D synergy is part of his broader investment synergy category.

We turn now to the criteria for assigning products to R & D projects, referring to the grouping of products launched both in different years and within a given year.

*Criterion 1.* Products launched in different years should, with minor exceptions, be grouped into different R & D projects. Thus each generation of a product should be regarded as a separate R & D project. The exceptions refer to accessories and peripherals (see criterion 2).

*Criterion 2.* If we want to emphasize differences in area variables between the products of a given year as explanations of differential performance, clear differences in market size or competition would call for separate projects, and vice versa. Accessories and peripherals, if not sufficiently standardized, will probably have to be combined with the main instrument or system with which they are associated.

*Criterion 3.* Up to a point, interproject synergy should be minimized to ensure that the outputs of each project depend as much as possible on its inputs and minimally on the inputs of other projects.[8] Strict application of this criterion will tend to reduce the number of projects in each year ("everything depends on everything") and will reduce the role of area-variable differences as an explanation of performance differences. Moreover, a grouping which minimizes intergroup marketing synergy does not necessarily minimize intergroup R & D synergy.

It is clear that there are a number of ways of grouping products into R & D projects, and that the best grouping for a particular year may depend on the products launched in previous (or future?) years. An analysis concerned with the growth of infant firms should stress criterion 1, while an analysis of comparative advantage should, I believe, emphasize criterion 2. There is much more to be said about this, but it seems preferable to wait before formulating a more general and precise theory of R & D project determination.

### Measuring and Analyzing R & D Project Performance through Time: An Illustration

We begin with a brief general description of the activities of the firm, based on the qualitative information whose publication was authorized at inception.

8. This was the main criterion used in Teubal, Arnon, and Trachtenberg 1976.

The firm started designing a variety of relatively simple electronic instruments within a particular area of technical expertise. Some of these failed technologically but a number succeeded and were launched, at first without significant commercial success. One of these products, however, incorporated a novel device that considerably improved its functional utility, especially for a user segment that previously had not generally purchased such a product. The instrument aroused considerable interest among users and although competitors launched similar instruments later on, our firm managed to earn a positive "direct" profit on both the first and subsequent generations of the product. This product was the basis for growth and profitability of the firm. It, however, became obsolete after a few years, and the firm was compelled to launch a new product of increased complexity, which it did after some delay and technological difficulties. The new product was not an instant success, unlike the previous one, and the direct profitability of the early generations was negative, but turned positive later on. The firm was late in launching the second product class, but it had gained enough reputation from the first product and from an important accessory to both the first and the second product (launched before actually launching the second product and compatible with the products of competitors) to be able to rapidly penetrate the market for the latter product. The set of nine projects selected include both product classes mentioned above, together with the corresponding accessories and peripherals, and two additional one-generation products which belong to the same overall market. It excludes other products developed by the firm in its early years (amounting to a small share of R & D and sales) and the R & D projects subsequently undertaken in connection with the launch of a completely new product class.

Table 6.1 lists the set of R & D projects of the case firm. The tens of products

Table 6.1. R & D productivity of the case firm's projects

| Project (i) | Product | | Base year | $\rho_i' = \rho_i/\rho_3$ [a] |
|---|---|---|---|---|
| 1 | I | | 1 | 0.35 |
| 2 | II | first generation | 1 | 0.00 |
| 3 | III | first generation | 2 | 1.00 |
| 4 | III | second generation | 3 | 1.85 |
| 5 | IV | | 4 | 2.89 |
| 6 | II | second generation | 5 | 0.33 |
| 7 | V | | 6 | 0.19 |
| 8 | II | third generation | 6 | 1.06 |
| 9 | II | fourth generation | 9 | 1.79 [b] |

[a] No significance attaches to the absolute values of these ratios. Their usefulness lies in showing the variation of project profitability within and across products.
[b] Underestimate.

developed and launched were grouped into nine projects spanning the ten years from the firm's inception, and belonging to five product classes (numbered I through V).

Products launched in different years are taken as separate projects provided that the sales of each are not dominantly dependent on the sales of the others. Sales of accessories/peripherals were almost totally dependent on the sales of other products although they contributed enormously to the sales of the latter. One group of accessories/peripherals was therefore included with a product II project, and another with a product III project.[9]

### The measure of performance

The measure of R & D project performance used is $\rho$, an R & D effectiveness (or productivity) ratio, namely, the ratio of the present value of real sales, $S$, of the project's products to the present value of real R & D costs, $R$; that is, $\rho = S/R$. The ratio $\rho$ is the project variable whose variation has to be explained (dependent variable). Its salient feature is that it is related to the direct rate of return and not to the project's total rate of return. A project's indirect contribution to subsequent projects (e.g., *via* sales) is not taken into account, but the firm-related variables responsible for it will explain a part of the difference between the $\rho$ of current and subsequent projects.[10]

### The appropriateness of $\rho$

It is natural to ask whether our project measure $\rho$ can be useful in general and a good proxy for project rate of return in particular. Several considerations suggest that it is.

First, in electronics, as in many high-technology industries, R & D is essential for survival, profitability, and growth. Second, in the electronics industry the share of R & D in the total investment (fixed costs) associated with innovation is large compared with other industries, such as chemicals. Therefore $\rho$ is presumably more relevant in electronics than in chemicals. Moreover, in electronics it is superior to the partial productivity index of other fixed items such as investment in physical capital;[11] in chemicals, where the investment out-

9. We recognize that there is no clear criterion for deciding whether a dominant share of the sales of a product is dependent on other products. A sufficient condition for "dominant dependence" is that there is no separately identifiable market for the product.

10. Note that later projects (e.g., a new generation) may also have an effect on the $\rho$ of prior projects (e.g., by reducing sales). This effect is probably not very important when product life is short.

11. Mansfield et al. (1971) calculate that the share of R & D in the total costs of launching approaches 50% in a sample of electronics innovations and that this exceeds the ratio found in a sample of chemical innovations. In general, the ratios are both area specific and country specific. They may be much lower in Israel, where R & D is relatively cheap and export marketing relatively expensive.

lays associated with the erection of the full-scale plant are relatively high, the opposite may be true. Third, investment in R & D on average precedes other investment items of the innovation process. They are therefore in some sense "more fixed" while the others may be regarded as "more variable." This justifies looking at the rate of return to project R & D (or alternatively, the ratio of sales to R & D) as an indicator of the rate of return to (total) project investment.

More specifically, the appropriateness of $\rho$ depends on what it is used for. One use is to describe the changing innovation performance of a particular firm within a particular area. For this purpose, the absolute level of the project rate of return may not be important: what matters is how it changes over time. The project $\rho$ constructed will be useful if its ranking corresponds to the ranking of project rates of return, and this will be so if the ratio of marketing to R & D expenditures and the markup coefficient do not vary (or do not vary substantially) across the sample.[12]

To illustrate, let $F_i$, $V_i$, and $S_i$ represent the present value of total fixed costs, total variable costs, and sales resulting from project $i$. Let $\pi$ be the present value of the stream of direct operating profits, $S_i - V_i$, per unit of fixed costs,

$$\pi_i = \frac{S_i - V_i}{F_i} = 1 + \frac{W_i}{F_i}, \tag{1}$$

where $W_i = -F_i + S_i - V_i$ is the present value of project $i$. We shall refer to $\pi_i$ as "project profitability." R & D costs are only one component of the total fixed costs of the innovation. Let $\alpha_i$ be the ratio of R & D to total fixed costs for project $i$, and $m_i$ the average markup of unit variable costs, i.e.,

$$R_i = \alpha_i F_i, \qquad S_i = (1 + m_i) V_i. \tag{2}$$

Using equations (2), equation (1) can be rewritten as

$$\pi_i = \frac{\alpha_i m_i}{1 + m_i} \frac{S_i}{R_i} = \frac{\alpha_i m_i}{1 + m_i} \rho_i. \tag{3}$$

Note that $\pi_i$ increases with $\alpha_i$, $m_i$ (given $\rho_i$), i.e., $\partial\pi_i/\partial\alpha_i$, $\partial\pi_i/\partial m_i > 0$. Therefore a sufficient condition for

$$\rho_j > \rho_i \rightarrow \pi_j > \pi_i \tag{4}$$

is $m_j \geq m_i$ and $\alpha_j \geq \alpha_i$. That is, if this condition holds, the R & D productivity index, $\rho_i$, and the project rate of profit, $\pi_i$, will give the same project ranking. In general the ranking of $\rho_i$ across a set of projects will correspond to the ranking of $\pi_i$ if the $m_i$ and $\alpha_i$ do not vary appreciably within the set. The

12. Alternatively, a calculation of the level of the project rate of return would require knowledge of these two parameters.

extent of variation of $\alpha_i$ and $m_i$ across projects will depend on the areas to which they belong and on other factors (e.g., whether or not they are successive generations of a given project).[13]

### The variation of $r$ and $\rho$

The project R & D productivity ratios, $\rho$, may be calculated by discounting real sales throughout product life and real R & D to the project base year, $t$, the year in which all or most of the R & D effort was undertaken. A positive real discount rate should be used throughout. Real discounted sales for a project is given by

$$S(t) = \sum_{k=0}^{n} \frac{s(t+k)(P_t/P_{t+k})}{(1+r)^k} \quad , \tag{5}$$

where $t$ is the base year, $k$ is number of years after base year, $s$ is nominal sales, and $P$ is the relevant price index. The real interest rate used in our calculations was 10%.

The variation in project R & D productivity relative to $\rho_3$ (i.e., $\rho_i' \equiv \rho_i/\rho_3$) is shown in table 6.1, where the projects are listed chronologically. The corresponding variation in $\pi_i'$ ($\equiv \pi_i/\rho_3$) is shown in table 6.2 for projects involving successive generations of a product class. The $\pi_i$ are computed from the $\rho_i$ under two alternative simplifying assumptions concerning $m_i$ and $\alpha_i$ (see note a to table 6.2; see also note a to table 6.1). Three striking features emerge from the tables.

1. Six out of the nine R & D projects shown are successive generations of two products. It should be noted that the very rapid transition from generation to generation and from one product class to another was accompanied by continuous expansion of the line of accessories and peripherals for both products and by continuous expansion of the number of models. Thus we observe both rapid change and rapid broadening of line, which is not typical of many other Israeli electronics firms during the period.

2. There are dramatic improvements in R & D productivity $\rho$ (or in $\pi$) *within* each product type—the ratios almost doubled for product III, while the low R & D productivity of the first two generations of product II is amply compensated for by the two later generations. Under our assumptions, a breakeven point is passed by both products (not shown in the table).

---

13. Across generations we observe an increase in $\rho$ (see table 6.1). While this does not necessarily imply an increase in $\pi$ and in project present value $W$, the latter has been confirmed by the qualitative evidence available. Moreover, an increase in the markup rate can also be justified on theoretical grounds (a lower markup in the first generation increases sales and thereby market feedbacks and reputation, which benefit the second generation).

Table 6.2. Project profitability ($\pi_i' \equiv \pi_i/\rho_3$) of different generations
of product classes III and II[a]

|  | Assumptions[b] | |
|---|---|---|
|  | (1) | (2) |
| *Product III* | | |
| 3. First generation | 0.16 | 0.25 |
| 4. Second generation | 0.31 | 0.46 |
| *Product II* | | |
| 1. First generation | 0.00 | 0.00 |
| 6. Second generation | 0.06 | 0.08 |
| 8. Third generation | 0.18 | 0.26 |
| 9. Fourth generation | 0.30 | 0.45 |

[a]See note to table 6.1.
[b]The ratio $\alpha_i$ is assumed to be 0.5 (following Mansfield, see footnote 11 above). The value of $m_i$ is arbitrarily set at $m_i = 0.5$ in column (1) and $m_i = 1$ in column (2). The figures are then calculated from table 6.1 according to eq. (3) in the text, with $\alpha_i m_i/(1 + m_i) = 1/6$ and $1/4$ for columns (1) and (2) respectively.

3. The projects show a gradual shift from one product class to another. The first generation of a new product may have a lower R & D productivity than the prevailing generation of the old. An example is the first product II to reach the market—its second generation's R & D productivity was below that of the prevailing generation of product III, 0.33 versus 1.85.

The trends in area variables and how the firm coped with them can be deduced from these characteristics.

*High rate of product obsolescence.* Independent evidence suggests that the generation life of products II and III was approximately three years. Moreover, new products that replace old ones in existing applications make their appearance from time to time.

*Successful adaptation of the firm.* The firm succeeded in adapting itself to these market and technology trends. However, the average trend of increasing project productivity over time is broken by temporary reductions which accompany entry into new product classes.

*Experience and learning are critical factors.* The experience and knowledge accumulated in previous projects endow the firm with the capacity to adapt itself to the high rate of product obsolescence and to the increasing complexity of the products in this area.[14] Without them, the launching of radically differ-

14. That the products are becoming more complex is confirmed by looking at the technologies incorporated into each product type and at the average price paid for products of the various generations and types.

ent and more complex products at short intervals and with increasing success would not have been possible (project productivity would tend to be less than unity and to decline over time). It follows that a high share of actual project effectiveness should be regarded as an *indirect* effect of prior projects and that this is increasingly so with the passage of time. This point requires considerable elaboration. The conclusion is reinforced by (1) direct evidence of strong R & D and marketing synergy between the various projects; (2) the fact that successful entry has become increasingly difficult, partly because it has become increasingly difficult to replace past experience by current investment; and (3) the possibilities of adapting old product classes to new applications are limited. Note that one cannot state unambiguously that market trends were favorable; despite the high rate of growth of the market, the market for existing products declined very rapidly.

### Estimating a project's indirect contribution

The indirect contribution of a project (say, project $j$) is its contribution to the profitability of subsequent projects; one—upward biased—estimate of this contribution is the difference between the profits actually received from the later projects and those that would have been received in the absence of project $j$. The discussion above suggested that in the absence of a fund of experience, goodwill, and a product line associated with prior projects, the effectiveness of the firm's projects would be low—probably less than unity and declining with the passage of time. In what follows we deal with the indirect contribution of $j$ to $j + 1$ (i.e., from one generation to the next generation only), and for simplicity we assume that it accounts for the *excess* of operating profits of $j + 1$ (present value, discounted at a positive real interest rate, $r$) over the fixed investments of $j + 1$.[15] This means that in the absence of project $j$ the firm could have earned a real rate of return of $r$ percent. Denoting by $\Pi_j$ the *total* profitability of project $j$ (direct and indirect), we get

$$\Pi_j = \pi_j + \frac{(S_{j+1} - V_{j+1}) - F_{j+1}}{F_j} \frac{P(t_j)}{P(t_{j+1})} \frac{1}{1+r}^{t_{j+1}-t_j}, \tag{6}$$

where the subscripts refer to the project in question. By multiplying and dividing the second term on the right-hand side by $F_{j+1}$, the real fixed cost of project $j + 1$ (discounted to its base year, $t_{j+1}$), we obtain

$$\Pi_j = \pi_j + (\pi_{j+1} - 1) \frac{F_{j+1}}{F_j} \beta, \tag{7}$$

15. Alternatively, under certain conditions one could assume that in the absence of $j$ the profitablity of $j + 1$ would be equal to the observed profitability of $j$. Both methods would understate the indirect contribution of $j$ (see below).

where $\beta$ is the last two factors in the second term of the right-hand side of equation (6). The $\beta$ coefficient discounts $F_{j+1}$ to the base period $t_j$ and deflates it to prices of $t_j$. This formula arbitrarily assumes that in the absence of project $j$ the profitability of project $j + 1$ is unity. Since the contribution of project $j$ to project $j + 1$ should not be negative in our context, formula (7) is thus applicable only when $\pi_{j+1} > 1$.

This history of the various projects provides us with concrete examples of indirect contributions flowing from one project to another. For example, the first product III project contributed substantially to the second *via* firm reputation, marketing arrangements, better understanding of user needs, R & D synergy, and other factors. Both product III projects also contributed *via* a line effect and some R & D and other synergy to the product II projects. Within the latter there are strong indirect contributions of the second and third projects to the third and fourth. Note that since it never reached the market, the first product II project could not contribute significantly to subsequent projects through firm-reputation and marketing-synergy effects. (R & D synergy with the next generation was probably also very low.)

The raw data underlying tables 6.1 and 6.2 show that one cannot always apply equation (7) to individual product II projects, since the underlying $\pi$ are less than unity. For illustrative purposes we therefore shall apply equation (7) to two calculations of indirect contribution: of the first to the second project of product III and of all projects related to product II (excluding the first genera-

Table 6.3. Direct and indirect profitability

|  | Assumption [a] | |
| --- | --- | --- |
|  | (1) | (2) |
| *Profitability of product III first generation* | | |
| 1. Direct ($\pi' = \pi/\rho_3$) | 0.16 | 0.25 |
| 2. Total: direct *plus* second generation profitability attributable to first generation ($\Pi' = \Pi/\rho_3$) | 0.30 | 0.63 |
| *Profitability of product II–related projects* [b] | | |
| 3. Direct ($\pi' = \pi/\rho_3$) | 0.17 | 0.24 |
| 4. Total: direct *plus* fourth generation profitability attributable to prior projects [b] ($\Pi' = \Pi/\rho_3$) | 0.24 | 0.46 |
| *Indirect profitability as percentage of total* | | |
| Product III, (line 2 − line 1)/line 2 | 46 | 61 |
| Product II, (line 4 − line 3)/line | | |
|  | 30 | 44 |

[a] The assumptions of columns (1) and (2) of table 6.2, respectively.

[b] Comprises the second and third product II projects and project IV. The base year was arbitrarily set at the base year of the latest of the projects in the set, the third product II project.

tion) to the successful fourth generation of product II. In this calculation, all the indirect benefits received by the latter are attributed to prior projects related to the same product class, as if the product III projects did not have any influence. In this respect, the result obtained should therefore be considered an overestimate.[16] The actual calculations are performed for $\pi'$ and $\Pi'$ (i.e., they are normalized by $\rho_3$).

Table 6.3 shows that the indirect contribution of the project or projects considered varies from 30% to 61% of total project profitability. This is a significant proportion indeed, especially considering the fact that a high real rate of interest (10%) was used throughout. The result is in line with the opinion of managers of high-technology firms, despite the crudeness of the method used to illustrate this point.[17]

## When will indirect project profitability be substantial?

Our purpose here is to give conditions under which the indirect contribution of early projects of the firm is expected to be high. These conditions refer to the environment, to the technological and/or market proximity between various projects, and to other factors. We assume throughout that the firm will either enter a new product class or remain active in its existing one.

*Nature of changes in the environment.* Changes in the environment which substantially raise the profitability of the new product class—provided it can be launched within a short period of time—while simultaneously reducing the profitability of the old class may lead to the above effect. A typical example is the exogenous availability of a new technology that induces a radical innovation with the following effects: it widens the overall market by increasing performance per unit cost, but at the same time, it *reduces* the market for existing products by rendering existing products obsolete. A profitability gap is thus created between firms that manage to switch swiftly to the new product class and those that do not.[18]

*Ability to launch a new product class.* The indirect contribution of the old product class will be high if the experience, goodwill, and infrastructure gathered provide a distinct advantage in launching the new product class, par-

---

16. The high real interest used, the nature of changes in the environment, and other qualitative evidence suggest that the figure arrived at is an underestimate of the real figure.

17. These orders of magnitude are probably not applicable to firms other than successful firms operating in dynamic product innovation areas.

18. This change in the environment is not unambiguously favorable or unfavorable to supplier firms; it depends on their capacity to adapt rapidly and efficiently to such a change. Changes of this kind should be contrasted with changes in the environment that are unambiguously favorable or unfavorable to firms in the area, for example, the natural growth of the market (e.g. due to population growth) which in principle would unambiguously benefit all the suppliers concerned.

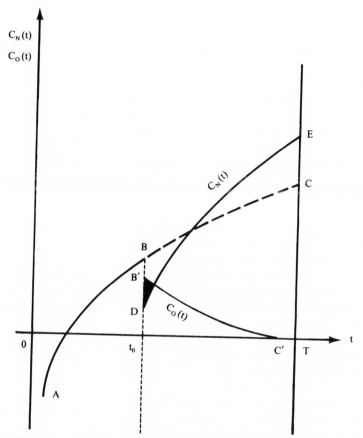

Fig. 6.1. Effects of an external shock on the profitability of the old and the new product class.

ticularly when sufficient experience is a necessary condition. This is likely to be the case if some of the technologies (and market experience) of the old class are relevant to the new class and if the latter is sufficiently complex to make it unlikely that a firm lacking prior experience could succeed.

A graphical representation of these conditions is shown in figure 6.1, where the vertical axis shows cash flows and the horizontal axis chronological time. $ABC$ represents cash flows associated with the old product class assuming an unchanged environment or a market growing at a natural rate. The positive slope of the curve (the counterpart of rising generation efficiency) may be due to endogenous firm experience or to growth in the overall market for the existing product class. At time $t_0$ there is a change in the environment of the kind described above—sales of the old product class would decline, a fact reflected in cash-flow curve $B'C'$. Sales of the new product class—the one associated with the radical innovation—could rise provided the firm is capable of launching it efficiently, and in time, the cash-flow curve in such a case is represented by $DE$.[19] If prior experience with the old product is necessary and sufficient for the launching of the new product class and for the firm to locate itself at $DE$ rather than at $B'C'$, then the value $V$ of this accumulated experience, etc., discounted back to $t_0$ is

$$V = \int_{t_o}^{T} [c_N(t) - c_O(t)] \exp[-r(t-t_o)] \, dt.$$

The quantity $T$ is the relevant time horizon, $r$ the real rate of interest, and $c_N$, $c_O$ are cash flows of the new and old product class respectively. The quantity $V$ (which is also the indirect profitability of the old product class) is roughly represented by the shaded area between the $DE$ and $B'C'$ curves.

We may summarize our results up to now. The indirect contribution of the firm's early projects is likely to be high if (1) there occurs an environmental shock that raises the expected profitability of the firm, provided it is capable of shifting swiftly and efficiently to a new product class (and lowers its expected profitability otherwise); and (2) the firm is capable of switching to the new product class thanks to the experience, knowledge, reputation, and infrastructure acquired from its early projects.[20]

### Summary and Implications

Our methodology for organizing data on project R & D input and output should make possible an orderly description of the R & D performance through time of a class of young, high-technology firms—one which will set the stage for subsequent analysis. This paper illustrates the approach for the case of a successful high-technology firm in the electronics area. The results obtained for this firm follow.

During the nine-year period studied, a wide variety of products were launched by the firm. Most of them belonged to product classes involving two

19. $DE$ may lie below $B'C'$ for a while after $t_0$, which simply expresses the presumption that shifting to a new product may require the firm to make substantial investments.

20. An alternative formula for calculating the value of a firm's experience is shown in Teubal 1984. It is valid for cases where prior experience—while helpful—is not necessary for entering a new product class. Such situations seem to be of greater applicability to metalworking areas characterized by a lower rate of product innovation relative to the dynamic electronics area covered here.

or more successive generations of products. Later product classes involved greater technological complexity than earlier ones.

Direct profitability increased across generations of a product class but decreased when shifting from an existing class to the first generation of a more complex, new product class. Early generations of a product class need not be directly profitable.

The pattern of rising profitability within a product class suggests that early generations contributed to the profitability of later generations (spin-off). This is confirmed by the qualitative evidence on R & D and market synergy across projects, and on the role of the experience, infrastructure, and reputation accumulated by the firm through time. Preliminary (and conservative) calculations suggest that the value of the indirect contribution of early generations to subsequent ones ranged from 50% to 200% of the direct profitability of the former.

Qualitative evidence also shows that early product classes contributed substantially to the direct profitability of later product classes. While the existence of interproduct synergy, a pool of common experience, and infrastructure, etc., has been documented in the individual project histories, their high economic value derived from the nature of changes in the competitive environment facing the firm. These changes may represent a threat or an opportunity to the firm; its high adaptability (derived from the activities associated with prior product classes) enabled effective launch of new product classes, thereby assuring its share in the expanding market opened up by the exogenous availability of new technology. This pattern is probably common to many successful firms in high-technology areas.

A final point is that the indirect profitability also depends on the R & D, tooling, and marketing investments associated with future products and projects. Substantial outlays were required in order to exploit the full benefit of current projects and this entailed substantial risks.

### Implications for the analysis of innovation performance

This paper does not undertake a microanalysis of commercial success and failure in innovation. Such an analysis requires a wider data base of R & D projects than that currently available and explicit consideration of relevant sets of explanatory variables, i.e., variables differentiating successful from failed innovations. For example, project SAPPHO[21] has focused on what may be termed *project execution* variables, such as the amount of marketing effort

21. See Science Policy Research Unit (1972); Rothwell et al. (1974); Rothwell and Teubal (1977).

and the size of the R & D team, that differentiate commercially successful from failed innovations among a set of instrument and chemical innovation pairs.[22] In Teubal, Arnon, and Trachtenberg 1976, on the other hand, emphasis was given to what have been termed *area* variables, that is, characteristics of markets and of technology that are exogenous to the innovating organization (e.g., market size).[23]

It is clear that the results of the paper are much more of a descriptive and methodological nature rather than analytical, in the sense of identifying factors explaining the commercial success of innovation. Specifically, we do not postulate criteria for project selection at the early stage of the firm's development, when it scanned various products lying within its technical expertise for one (or more) which might provide the base for its growth and profitability. What it does is to show that once such a product was found, generations characterized by increasing direct profitability appeared and that the "spin-off" is high from generation to generation (and from product class to product class).[24] This paper does, however, give a dynamic perspective to the problem of performance, one which distinguishes direct profitability from total profitability.[25] The problem for the firm is to choose a project with high *total* profitability and

22. The SAPPHO study shows that successful instrument innovations within a set of 21 pairs (each pair having a success and a failure) were associated with a higher level of marketing expenditures, a better understanding of user needs, and better communications with the outside technical community relative to the corresponding failed innovations. This result was obtained from paired comparisons, i.e., comparing the successful innovation in a pair with its failed counterpart and not with other failures. Within a sample of 22 chemical pairs, the successful innovation involved a greater R & D team, shorter development times, an earlier launching date, and greater authority and seniority of management relative to the failed innovations. It should be noted that none of these factors significantly differentiated successful from failed innovations in instruments (the *opposite* result was obtained for some of these).

23. For the whole universe of Israeli biomedical electronic innovations till early 1974, all "successes" involved improvements in product classes whose use was already standardized. This means, first, that users of the innovations launched by the firms of the sample had already defined needs (area variable) and, second, that these innovations improved upon their "preferred" class of products rather than involving the launching of completely new product classes (product characteristic). Moreover, within this group of innovations, successes relative to failures emphasized improvements in performance rather than reductions in price. Most failures, on the other hand, involved the launching of completely new product classes, either to users without a preferred class (undefined needs) or to users with a preferred class.

24. In a sense, within the sample of innovations studied, firm endogenous factors "explain" the increasing direct profitability across generations and the fact that the profitability of the new product class launched is substantially higher than the profitability of the R & D projects that would be undertaken in the absence of these factors.

25. The quantitative and continuous measure of performance proposed in this paper is important in this connection, e.g. when the generations of a product are all "successes" (positive "direct" profitability) but the degree of success varies dramatically from one generation to another.

this need not imply high or even positive *direct* profitability. Existing studies of success and failure do not make these distinctions and therefore their conclusions may not be relevant when spin-offs are high. Specifically, it may be possible to spell out formally conditions which imply both negative direct profitability in early generations and positive total profitability for the corresponding product class.[26]

## The economic value of indigenous firm spin-offs

We have seen that significant spin-offs were generated along the development path of a successful Israeli electronics firm and that this may have implications for an analysis of innovation performance. The economic significance of these "indirect" effects has also been documented in a study of the learning processes within a group of Brazilian and Argentinian capital-goods firms (Teubal 1984). What can we say, from this and other work, about the factors determining the economic value of these spin-offs? When are they likely to be high?

In general, we may say that the economic value of the endogenous spin-offs within a firm depends on the magnitude of the experience, knowledge, and reputation accumulated, on the nature of the environment and changes in it, and on the actions of the firm concerning future products/projects. Some specific points follow.

Changes in the environment need not be unambiguously favorable or unfavorable to the firm, as when a high rate of growth of the overall market is associated with increasing product complexity and a high rate of product obsolescence. Under these circumstances, the effects depend critically on the firm's adaptability, and this largely depends on the magnitude and type of experience, knowledge, and reputation acquired from earlier projects of the firm. When the firm is highly adaptable, the effect of the change in the environment will be relatively favorable, and correspondingly the indirect profitability of earlier projects will be high.

The magnitude and structure of the complementary investments required varied considerably from case to case. The electronics firm, for example, had to invest considerable amounts in R & D and in marketing; its investments in equipment for manufacturing and for quality control were apparently relatively smaller than those required for metalworking capital-goods firms who shifted to the production of heavier goods or to the production of goods with more exacting specifications.[27]

Government policy may help (e.g., by providing investment subsidies) or

---

26. These conditions are currently under investigation.
27. Additional research is required to determine the generality of this statement.

hinder (e.g., by a systematic overvaluation of the currency) the exploitation of potential spin-offs.[28]

### Spin-off mechanisms

Our purpose is not only to assert that endogenous firm spin-offs in our electronics firm were significant; it is also to state something about the nature of these spin-offs and differences between these and those found in firms belonging to other industries. In what follows I summarize some of the conclusions obtained up to now.

The most significant spin-off mechanisms operating in our young electronics firm were associated with R & D, marketing/market feedbacks, and firm reputation. Thus, it would seem to be insufficient to represent the stock of knowledge of the firm by cumulated past R & D exclusively, as has been the tradition in quantitative studies of productivity; the other two factors are extremely important as well.[29] There is some evidence that the spin-offs from projects not culminating in product launch are considerably lower than those for projects leading to products actually marketed.[30]

The relative importance of the various spin-off mechanisms seems to vary with the areas and firms being considered. For example, among the 11 mechanisms identified in the sample of capital-goods firms mentioned above, the most important and pervasive was the accumulation of manufacturing abilities (including plant-operating experience and experience in quality control). These enabled the firms to initiate the production of more complex products associated with emerging market trends. The role of R & D, unlike the situation in a highly dynamic area like electronics, is insignificant, especially at the early stages in the development of these firms (Teubal 1984). In oil refining, on the other hand, available evidence suggests that the introduction of a new process lays the groundwork for a succession of minor improvements based on operating experience and possibly R & D.[31]

---

28. When the direct profitability of the early projects is low or negative, government support will be justified only if the indirect social profitability of the projects is high and when capital markets are imperfect.

29. Specifically, market feedbacks from the first generation of a product have the effect of reducing the R & D "target uncertainty" of the second generation. Studies on productivity growth of high-technology firms have focused on estimating the R & D coefficient of company-level production functions. See Griliches (1980), where a study is made of over 1000 United States R & D performing corporations.

30. This statement may have considerable implications for the innovation strategy of young firms in electronic instruments areas.

31. See Enos 1958, in which the $\alpha$ and $\beta$ stages of a new process are defined. In oil refining, the economic significance of the minor improvements (occurring at the $\beta$ stage) exceeds that of the implementation of the new process (during the $\alpha$ stage).

The upshot of the various studies reported here is that the spin-off mechanisms operating in firms are, to a considerable extent, sector specific.

There are several additional points worth mentioning: Not all of the spin-off mechanisms are "technological" or refer to technological knowledge, and the accumulation of operating or production experience need not represent technological change. Technologically sophisticated firms may obtain significant nontechnical spin-offs from current activity, such as market feedbacks and firm reputation effects. Firm reputation effects may be critical in areas like pacemakers, aircraft, and computers for process control, where the costs of failure are extremely high; they are also more important for new firms than for established ones.

### References

Ansoff, H. Igor. 1968. *Corporate Strategy*. Harmondsworth: Penguin Books.

Enos, John L. 1958. "A Measure of the Rate of Technological Progress in the Petroleum Refining Industry." *Journal of Industrial Economics* (June), pp. 180–97.

Griliches, Zvi. 1980. "Returns to Research and Development Expenditures in the Private Sector." In *New Developments in Productivity Measurement and Analysis*, edited by J. Kendrick and B. Vaccara. Chicago: University of Chicago Press.

Mansfield, Edwin, et al. 1971. *Research and Innovation in the Modern Corporation*. New York: Norton.

Mansfield, Edwin, et al. 1977. "Social and Private Rates of Return from Industrial Innovations." *Quarterly Journal of Economics* 91:221–40.

Myers, Sumner, and Donald G. Marquis. 1969. *Successful Industrial Innovations*. National Science Foundation NSF 69-17. Washington, D.C.

Rothwell, R., C. Freeman, A. Horsley, V. T. P. Jervis, A. B. Robertson, and J. Townsend. 1974. "SAPPHO Updated—Project SAPPHO Phase II." *Research Policy* 3:258–91.

Rothwell, R., and M. Teubal. 1977. SAPPHO Revisited: A Reappraisal of the SAPPHO Data." In *Innovation, Economic Change and Technology Policies*, edited by K. Stroetmann. Basel and Stuttgart: Birkhauser.

Science Policy Research Unit, University of Sussex. 1972. *Success and Failure in Industrial Innovation: Report on Project SAPPHO*. London: Centre for the Study of Industrial Innovation.

The background of this paper is a series of case studies on Israeli electronics firms, conducted over a number of years with Manual Trajtenberg, without whose collaboration this paper would not have been possible. I appreciate the cooperation of the firms involved. The paper has also benefited greatly from the comments and editorial help of Susanne Freund and from the collaboration of Shaul Lach. The encouragement and comments of A. Kalisky, H. Arbell, and Z. Adelman of the Office of the Chief Scientist, and of Y. Ben-Porath and two anonymous referees are greatly appreciated. I am indebted to the Office of the Chief Scientist at the Ministry of Industry, Commerce, and Tourism for financial support.

Teubal, Morris. 1984. "The Role of Technological Learning in the Exports of Manu-factured Goods: The Case of Selected Capital Goods in Brazil." *World Develop-ment* 12:849–65. (Chap. 5 in this volume.)

Teubal, Morris, Naftali Arnon, and Manuel Trachtenberg. 1976. "Performance in Innovation in the Israeli Electronics Industry: A Case Study of Biomedical Elec-tronics Instrumentation." *Research Policy* 5:354–79. (Chap. 1 in this volume.)

Utterback, James M. 1975. "Successful Industrial Innovations: A Multivariate Analy-sis." *Decision Sciences* 6:65–77.

# 7 The Accumulation of Intangibles by High-Technology Firms

## Morris Teubal

### Introduction

Innovation and technical change have long been recognized as significant factors in the growth of firms and industries, and attempts are being made to incorporate these phenomena into theory and empirical analysis. This paper is an attempt to analyze empirically the significance of innovation. For our purposes, it is useful to distinguish quantitative efforts to measure the contribution to productivity growth from microeconomic case studies of individual firms or sectors. The standard procedure in the former is to posit for every sector or firm included in the analysis a relationship between explicit inputs to innovative activity (R & D expenditures) and output, profits, or productivity (see Griliches 1980). This procedure ignores the variety of mechanisms for the accumulation of intangibles or for the creation of capabilities, of which knowledge from R & D is just one example. It also ignores the variety of ways in which innovation affects growth. Both have been illustrated by the increasing number of case studies of firms and industries from a number of different countries, sectors, and contexts. Studies of innovation performance, such as project SAPPHO (Science Policy Research Unit 1972; Rothwell et al. 1974), have shown that the factors separating the successful innovations within a set of instrument innovations from the failed ones are to some extent different from those separating successful chemical innovations from failed ones. The relative importance of factors relating to marketing, understanding of user needs, and market feedbacks was found to be higher in the former set of innovations than in the latter. These differences may indicate that different industries accumulate different kinds of intangibles. Another tradition of research—one that focuses on the growth of firms through time—shows that operating experience, rather than explicit R & D, is a central factor explaining the increased efficiency of new processes in industries such as petroleum refining and steel (Enos 1962; Hollander 1965; Maxwell 1977; Katz et al. 1978;

Dahlman and Valadares Fonseca 1978).[1] In metalworking firms, on the other hand, a crucial mechanism explaining the learning and technological capacitation process is the accumulation of manufacturing abilities, with design capability and R & D appearing only at a later stage of the process (Teubal 1984; Castano, Katz, and Navajas 1981).[2] It follows that it may not be appropriate to use R & D as one of the intangibles or the only intangible for all branches of industry.[3]

This paper makes a contribution toward the conceptualization of the various mechanisms of intangibles accumulation that have been identified in various industrial branches. It is hoped in this way to strengthen the links between case studies and more general quantitative attempts to measure the contribution of innovation to economic growth. In addition, it may help us in understanding the indirect contribution of production, innovation, and investment in a variety of branches. More specifically, the purpose is (1) to distinguish patterns observed in process industries and in metalworking from those observed in dynamic electronics areas, (2) to illustrate the complex nature of intangibles accumulation within the latter, and (3) to suggest some ways of representing the various patterns. Throughout the discussion a distinction will be made between the direct profitability of activities such as investment and innovation, and their indirect profitability. The latter results in part from the accumulation of intangibles, including the effect of activities at any given time, $t$, in changing the profitability of future activities.

## Intangibles Accumulation in Process and Metalworking Industries

The pattern of technical change and intangibles accumulation that has most often been considered in case studies relates to the process industries and to metalworking. This will be reviewed as a background to the description of the pattern observed in some electronic instruments firms.

### Operating experience in process industries

The studies by Hollander (1965) and Enos (1962) document the achievement of increased productivity and reduced unit production costs in rayon production and in petroleum refining respectively. Hollander investigated the history

---

1. R & D may, however, still be critical for the launching of new processes.

2. Firm reputation effects were also found to be very important within the Brazilian capital-goods firms studied in Teubal 1984.

3. Another criticism of the production function approach is that it ignores the relationship (1) between intangibles accumulation and the accumulation of physical capital and (2) between economic growth and innovation. The latter has been emphasized since the work of Schmookler 1966.

of Du Pont's rayon plants for periods of between 14 and 23 years, during which time unit cost declined by yearly rates of between 2.3% and 4.9%. An attempt was made to separate the effects of technical change from those of economies of scale (for example, the spreading of fixed costs resulting from the duplication of plants and other factors). Technical changes were classified, first, as major or minor (according to difficulty and the amount of investment required) and, second, into those changes which reduced costs at existing levels of output (for example, waste reductions) and those where cost reductions depended on increases in output. The latter cost reductions derive either from the stretching of capacity (termed indirect technical change, as might be achieved by increasing spinning speeds of existing machinery) or from the introduction of new machinery. Hollander's analysis showed that technical change was of overwhelming importance in explaining the observed reductions in cost, with minor changes accounting, in all rayon plants except one, for at least 75% of all the reductions in cost due to this factor. Moreover, most of the changes were of the capacity-stretching type; they originated in current operations and involved the cooperation of technical assistance to production groups rather than the central R & D laboratory. Additional microeconomic work confirmed the importance of operating experience and minor improvements in raising productivity in the context of rayon plants in Argentina and in steel plants (Katz et al. 1978; Dahlman and Valadares Fonseca 1978; Maxwell 1977). These studies showed that this experience led to substantial capacity stretching—minor improvements whose main objective is increased capacity rather than direct reductions in cost. They also suggest that capacity stretching was an efficient alternative to conventional investment, although the firms involved turned to it only when forced by external circumstances.[4] In addition, a study of a big Brazilian producer of welded pipe confirms an additional fact found from the steel plant studies, namely, that operating experience was a central determinant of the increased efficiency of the firm in its subsequent investment program. The mechanisms involved include experience with the layout of machinery, with the operation of existing machinery, which permitted more efficient designs, and the greater capability of predicting the times and costs of erecting new plants.[5]

The above pattern of intangibles accumulation from investment and plant operation has implications for determining both the direct and the indirect

4. For example, by an unexpected financial constraint blocking the way to conventional expansion.

5. See Teubal 1984. Capacity stretching of an existing plant would be one outcome of operating experience, one that is (1) disembodied and (2) can be incorporated into the existing plant. Part of the remaining operating experience would benefit only new plants.

profitability of activities in the process industries. Thus, the direct economic return to investments at $t$ (cumulated operating profits net of investment costs) may have to include the effects of operating experience in reducing costs via minor improvements and capacity stretching. In a very important sense, all of these benefits should be attributed to the original investment since the original investment is the base which sustains the string of improvements.[6] In addition to the direct return, there is an indirect return to investing at $t$. This derives from the lower capital requirements, adjustment delay (and gestation period), and risk associated with subsequent investment programs. They result both from the planning and execution of the original investment and from the operating experience accumulated.

### The case of a successful steel plant in Brazil

The study of the Usiminas plant by Dahlman and Valadares Fonseca (1978) provides very interesting material on intangibles accumulation within this very successful steel firm. The history of the firm could, in principle, be described as a succession of investment spurts or programs where the planning, execution, and operation of a program led to the accumulation of intangibles which benefited subsequent programs. The original plant had a design capacity of 500,000 tons per annum and was planned and installed almost wholly by Japanese minority partners in the firm. Production began in 1963, and after three years the design capacity output was achieved. Between 1966 and 1972 output increased continuously, and thanks to a succession of minor technical changes and improvements, reached a level of 1,200,000 tons per annum—a stretching of 140% beyond nominal capacity. The first expansion for an additional 1,200,000 tons (the second investment spurt) was completed in 1973. The expansion benefited principally from operating experience in the original plant, and the Brazilians acquired substantial engineering capabilities from their participation in the planning and execution of this project. In fact, they performed about 40% of the engineering. This experience was very important for the second expansion (the third investment spurt), which raised capacity by an additional 1,100,000 tons per annum and for which planning and execution were the total responsibility of the Brazilians. The company subsequently sold engineering services to other steel plants both within Brazil and abroad. The data appearing in the case study are not sufficient to calculate the direct rate of return for each investment program, and certainly not to calculate the value of the indirect contribution of each investment program to subsequent

6. In other words, the direct return to investment at $t$ should include a portion of the indirect return, or spin-off, from current production. If operating experience does not translate automatically to costs—if other investments in skill, etc., are required—then only a part of the operating profits resulting from lower costs should be attributed to the other (complementary) investments.

ones. There is, however, substantial information on the nature and mechanisms of intangibles accumulation.

*Operating experience.* A lot of information on the original installation and how it affected the first expansion program is available. The firm gradually learned how to operate the equipment and, in particular, to select and prepare the available raw materials, a process requiring considerable experimentation, design changes in the machinery, and the addition of auxiliary equipment. It also learned how to stretch the capacity of the plant, a process which also involved experimentation and design changes, although actual investments were low (the original plant with nominal capacity of half a million tons costs $261 million, while the stretching to 1.2 million tons was estimated to have cost only $40 million). Specifically, operating experience brought better specification of equipment: the new sinter machine, for example, incorporated screening and roll crushers to control sizes of the mineral ore entering the machine, while the volume of the new blast furnace was specified on the basis of the firm's experience with the selection and preparation of raw materials. Changes in auxiliary equipment and in refractory linings, some of them associated with capacity stretching in the original plant, were also specified. Similar changes were made in the specifications of the new steel shop, which incorporated both the results of past experience and exogenously available improvements. The result of all this was the erection of a more efficient plant at less cost. Both unit capital costs and unit operating costs (materials, labor, and energy) of the second investment program were lower than they would have been in the absence of the original investment. Better layout was also stated to have resulted from past experience.

*Acquisition of engineering capabilities.* This was particularly striking during the first expansion plan, and it gave the firm total independence from a foreign consultant during the second expansion plan. It should be noted that in the same way that the acquisition and utilization of operating experience required explicit investments and organizational changes (such as establishing and staffing the Industrial Engineering Department and support-oriented Research Centre), the acquisition of an engineering capability required working in close collaboration with a foreign consultant and investing heavily in know-how and in training.[7] In 1970 a separate Superintendency of Development was established, and by 1975 it embraced a number of departments (Process Engineering, Basic Engineering, Equipment Engineering, etc.). This case seems to show an interconnected sequence of intangibles accumulation, starting with operating experience in a broad sense—including knowledge of materials, process

7. Maxwell (1977) mentions how another aspect of operations—preventive maintenance—enhanced the design capabilities of the Argentinian steel firm he studied. The overall picture of intangibles accumulation of this firm resemble that of the firm discussed here.

and quality control, maintenance procedures, etc.—followed by equipment design and engineering investment capabilities, and even equipment manufacturing capability.

### Manufacturing abilities in metalworking

The original learning-curve concept relates unit costs of a particular product to the accumulated output of that same product. This has been shown to be relevant in the production of a wide variety of items (metallic and others) from airframes to integrated circuits. However, case studies of individual metalworking plants show that in addition to the above simple learning, firms learn to produce increasingly complex or sophisticated products (e.g., products with stricter specifications and/or greater weight). This phenomenon of qualitative learning is extremely important for understanding the growth of infant capital-goods firms in developing countries, and in particular for understanding the emergence of an export capability (see Teubal 1984).

There are several points concerning qualitative learning that are worth mentioning. While operating experience involving the plant as a whole dominates in the process industries, the manufacturing abilities being accumulated in metalworking firms refer to individual operations or processing steps common to a wide variety of products (Rosenberg 1976). They involve greater skill in performing both individual operations and an increasing array of operations. It follows that the accumulation of these abilities is the basis for product diversification, and in particular for the transition from simple to complex products. (The counterpart in the process industries is the increasing ability to plan and execute complex investment programs.) For example, the welding of overhead cranes in a heavy-capital-goods Argentinian firm contributed to the capability of welding nuclear power components of much stricter specifications —and to obtaining orders for these products.[8]

Complementary investments required to enter the new product lines may be very substantial when the new lines involve heavier or more sophisticated products. They probably involve more expensive equipment for quality control as well as new production machinery. Thus, in contrast to simple learning, intangibles accumulation is not a pure spin-off in this case. The accumulation of manufacturing abilities sets the base and increases the incentive for acquiring a design capability first in a relatively narrow range of products. This need not mean R & D and prototype testing but rather a capability for tailoring a particular product to specific users. (The equivalent design capability in process industries may refer to process machinery or to the capability of planning or

8. Product diversification also takes place and may even be easier within a particular product class. This particular firm launched a series of increasingly heavy overhead cranes (starting with one for in-house use), including units for ports, the steel industry, and hydroelectric dams.

executing investments.) Formal R & D may come later on, sometimes as a spin-off of quality control.

### The case of a successful lathes producer

The work by Castano, Katz, and Navajas (1981) is an interesting case of the kind of study that may lead to a better understanding of the main issue of this paper. The authors divide the history of a successful Argentine machine-tool producer into a number of stages, each one characterized by its organizational structure and by a set of product and process innovations. Although the study does not directly focus on the process of intangibles accumulation, it provides significant information on some aspects of this process, especially those related to the growth of a design capability.

During stage 1 (1945–1960) the firm was a family-owned artisan shop devoted to producing foreign-designed lathes for the domestic market (for example, universal machines for repair and maintenance shops). There was no real design effort, merely the copying of foreign models, although these models increased in complexity over time. While the first item was copied in an artisan way (without extensive use of drawings and plans), the last item copied—the Ursus lathe (1958)—required the services of a professional design engineer.

The second stage (1960–1965) began with absorption of the shop into a larger foreign-owned group which had only recently set up a small shop of its own for the supply of special lathes to the emerging automobile and parts producer market. That small shop's manager, a German engineer with extensive design experience, became the head of the new company and formally set up the design department of the firm. During this period of very rapid growth, the firm copied or produced under license a few models of special lathes for the new markets. Toward the end of the period, and in response to a decline in these markets, it launched its first product wholly of its own design—the highly successful universal T-190 lathe. This product pushed the firm to a leadership position within the Argentine machine-tool sector. It incorporated improved features, such as increased weight to sustain the higher speeds of a stronger cutting tool. Mastery of new techniques, such as heat treatment and the rectification of gears, was required. Investments in equipment, training efforts, and organizational changes were critical to ensure success of the T-190 (firm output tripled during the period). Although the authors state that previous experience in copying the Ursus was important for designing the T-190, they do not focus on the role of preexisting production skills (and possibly reputation and knowledge of markets) in enabling the firm to diversify.

During the third period (1965–1969), the local market slackened and the firm responded by adapting the basic T-190 model for a variety of user segments and by widening and improving the line of special machinery. The skills and infrastructure developed for the T-190 must have been very useful for this new effort of diversification, but more information on this is needed. The study focuses, however, on the complementary efforts undertaken to ensure that the firm maintained its previous levels of output and productivity throughout the crisis, given the wide variety of products being manufactured. These efforts included the establishment and development of separate departments of methods, design, and production planning and control and the employment of more people not directly engaged in production (including technicians and engineers).

This interesting case shows that it may be possible to structure the history of the firm as a sequence, or tree, of products or product lines (Ursus, T-190, specialized lathes) linked by production, design, and possibly other types of experience (spin-offs). In addition, a critical independent role should be attributed to discrete, non-project-specific investments, to changes in organization, and to incorporation of new technical personnel. Such a firm profile may shed new light on the role of experience in productivity growth and may be more useful, given the importance of new products, than a profile based on investment spurts, such as the one suggested for steel firms.

### Intangibles Accumulation in Electronics

The available evidence suggests that the learning process within electronics firms differs considerably from those suggested above for the process and metalworking industries. The description that follows is based on the R & D histories of a series of electronics firms during the 1970s, and in particular on the development of a very successful Israeli instruments firm. A full report on the latter can be found in Teubal 1982: only some aspects of the firm's development will be considered here. The R & D history was expressed in terms of a series of R & D projects representing successive generations of a given product, transitions to more complex substitute products, and diversification to other products. The qualitative information suggested very strongly that the profitability of a project at time $t$ depended considerably on the projects undertaken prior to $t$, and that the magnitude of this effect depended on changes in the environment (for example, exogenous availability of new technology such as new electronic components). In the mechanisms involved, production experience did not figure prominently, and R & D, although important, could only serve doubtfully as a proxy for the multitude of mechanisms involved.

#### The set of projects and their direct profitability

The products launched by the firm in the area covered by the case study were grouped into nine R & D projects belonging to a total of five product classes.[9]

---

9. These were the main areas of the firm during its first 10 years of existence.

Table 7.1. R & D profitability of the case firm's projects

| Project (i) | Product | | Base year | $\rho_i' = \rho_i/\rho_3$ [a] |
|---|---|---|---|---|
| 1 | I | | 1 | 0.35 |
| 2 | II | first generation | 1 | 0.00 |
| 3 | III | first generation | 2 | 1.00 |
| 4 | III | second generation | 3 | 1.85 |
| 5 | IV | | 4 | 2.89 |
| 6 | II | second generation | 5 | 0.33 |
| 7 | V | | 6 | 0.19 |
| 8 | II | third generation | 6 | 1.06 |
| 9 | II | fourth generation | 9 | 1.79 [b] |

[a] No significance attaches to the absolute values of these ratios. Their usefulness lies in showing the variation of project profitability within and across products.
[b] Underestimate.

There is more than one project for product classes II and III—four and two projects respectively—each one corresponding to a particular generation of that class. During its first year of existence, the firm was involved in three R & D projects, and only one was directly profitable—the first generation of product class III. The first generation of product II was a technical failure and was never launched. The list of projects and their direct profitability is shown in table 7.1.[10] The direct profitability measures $\rho_i'$ are based on the ratio of the present value of sales to the present value of R & D expenditures, with the base period being the year in which most of the R & D was undertaken.

Each point of figure 7.1 shows the year in which most of the R & D of a particular project was undertaken (in most cases, the year of launch of the main product included in the project) and the profitability of that project. The solid line connects successive R & D projects, while the broken lines connect the various generations of products III and II. The figure clearly shows that while the direct profitability of successive generations of a product increased, the shift from one product to another may be associated with a reduction in profitability. Products IV and V represent efforts at diversification on the part of the firm, while product I groups a line of products tried by the firm at its inception and later abandoned in favor of more successful products. Products II and III involve a principal instrument and a series of optional accessories and peripherals, launched at various intervals of time to some extent common to both product classes. Although both II and III served somewhat different users at an early stage, technological developments eventually gave II a deci-

10. Further details on the grouping of products into projects and on measuring the direct profitability of the latter can be found in Teubal 1982.

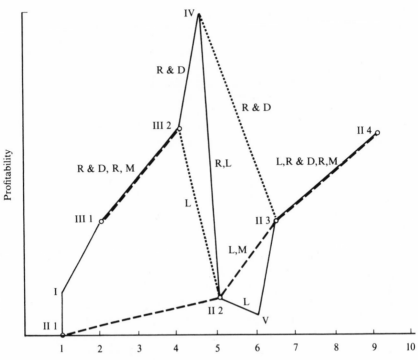

Fig. 7.1. Project profitability and interrelationships (instruments firm).

sive advantage over III for all users. The first generation of product II may thus be regarded as a premature attempt by the firm to innovate in this area— one that was suspended prior to attaining technical success. In general terms, successive generations of a product were more complex, commanded a higher price, and involved higher R & D costs. Instruments belonging to II were also, generally speaking, more complex than those belonging to III.

## The indirect profitability of projects

The pattern of rising direct profitability across generations and the qualitative evidence suggested that early generations contributed to the profitability of later generations. In other words, the former's total profitability exceeded direct profitability. In addition, qualitative evidence on projects and product classes shows that product classes launched earlier contributed to the profitability of product classes launched later on. This contribution was significant because the firm managed to adapt itself efficiently to exogenous changes in the environment, such as the availability of new components, which reduced the profitability of old generations/product classes and increased the profitability of new generations/product classes.[11] What were the main factors associated with a project which benefited subsequent projects?[12]

We may distinguish two main means by which a project benefits subsequent projects: through intangibles and capability based on physical capital, both accumulated while executing the earlier project. The qualitative evidence is overwhelming in indicating the importance of the former group of factors and the relative unimportance of the latter. This is probably much more typical in young electronic instruments firms than in metalworking firms. Young electronics firms usually start investing in plant and equipment once a successful prototype has been tested and strong demand signals suggest that it may have a market; in particular, an order to supply a requisite amount of equipment is sometimes a requirement for investment, and this requires a pool of marketing and technical knowledge. Also, products are not standard, so cost-effectiveness based on integrated and massive production processes is not a central requirement for success.[13]

What follows will concentrate on four types of intangibles flowing from one project to another: useful technological knowledge, or R & D synergy (simply termed R & D); useful knowledge obtained from marketing or from user feedbacks ($M$); a line effect ($L$), which may be one-way when the earlier project benefits the later project, or two-way when, in addition, existence of the later project increases sales of the earlier project; and firm reputation effects ($R$), which generally, but not exclusively, result from past sales. In our firm, R & D includes instances where a whole subassembly developed for a particular generation is the base upon which improvements on it are incorporated into subsequent generations. $M$ includes instances where user feedback or marketing efforts have shown the desirability of changing product specifications or the desirability or possibility of offering additional products. $L$ will exist whenever old customers of the firm are more likely than new customers to purchase an "add-up" to the instrument purchased in the past or another instrument which complements the one purchased in the past in some way (due to familiarity in use and in maintenance, ease in interfacing, etc.), or when new customers or dealers have a strong preference for purchasing from a

11. Changes in the environment were thus not unambiguously favorable or unfavorable to the firm.

12. In Teubal (1982), I mentioned experience, infrastructure, and firm reputation. My purpose here is to be more specific, both with respect to the various components of the pool of intangibles and with respect to the individual projects which benefit from them. The main source of information was the 220-page file of the firm, which contains a description of each project. The information is not necessarily complete, since the interviews were not closed ones systematically covering the various projects; nevertheless, I do believe that they indicate the main influences at work.

13. Production is largely based on relatively easy assembly operations, and involves short runs.

firm offering a complete line of products rather than from firms offering individual products only.[14] Finally, $R$ is generally acquired from past sales of similar or related products which have satisfied customers from a functional point of view, but it also includes an instance where the firm acquired a reputation for technical excellence and sales potential in the eyes of a prospective sales agent.

### The pattern of relationships observed

A "minimum" pattern of the more pronounced project interrelationships is shown in figure 7.1, where the symbols for the various effects are located over the segments connecting two projects. Most of the symbols are located on segments joining two consecutive projects. The important exceptions are shown by the dashed lines drawn from NW to SE and connecting III2 and II2 (the second generations of products III and II) and IV and II3 (product IV and the third generation of product II). The main points are as follows. (1) There are a number of types of intangibles accumulation, only one of which is R & D. The accumulation of production experience across projects did not seem to be very important, although the experience curve operated within individual products. (2) Relevant intangibles benefited both subsequent generations of a product class and also the design and sale of new product classes. Thus product obsolescence was much stronger than intangibles obsolescence. Even when indirect benefits exist potentially, their actual realization depends on the capacity of the firm to shift successfully to new products and thereby assure its own growth and profitability. This requires, in addition to the intangibles accumulated, significant investments, principally in R & D and in marketing. (3) The indirect benefits of the project not leading to sales (the first generation of product II) were low or nonexistent, even with respect to R & D. In general, it seems that little information on the targets of R & D for subsequent generations can be accumulated without actual sales and user feedback.

## Issues and Problems

### Intangibles from inputs or outputs

We have shown that a number of mechanisms for intangibles accumulation in industry operate, and it may be useful to summarize them by distinguishing between those which flow from the application of conventional inputs to par-

14. A two-way $L$-effect also occurs within projects and not only across projects. This is because optional accessories and peripherals are grouped together with the main instruments with which they are associated.

ticular activities or projects and those which flow from R & D (input to particular innovation projects) or from marketing. This picture is evidently too simplistic because it abstracts from critical intangibles obtained from users —market feedbacks—an important part of which depends crucially on sales, that is, on outputs of innovation projects.[15] Similarly, the experience and knowledge relevant to investments depend on both inputs (participation of a firm's personnel in the planning and execution of investments) and outputs (operating experience of the new plant). There are, however, some intangibles which may be associated with either input or output—for example, production skills and firm reputation, which generally depend on inputs and on outputs (successful sales of new products) respectively. Finally, several studies make reference to the accumulation of design skills in firms, particularly in the metal-working and machinery sectors, rather than the accumulation of technical knowledge. What factors explain this process? It may be tempting to distinguish between pure design skills—a capability to design products of increased technical performance relative to existing products—and a capability to produce designs which improve the utility or service provided to users (relative to cost) and which can be manufactured by the firm concerned. The former does not necessarily lead to the latter, as is already well known from innovation studies.[16] Although pure design skills may arise from past experience with purely technical activities and from current R & D, effective design skills—those which in principle can lead to commercially successful designs—depend also on market knowledge and on knowledge about manufacturing capabilities.[17]

15. The evidence on this is very extensive, especially for the machinery, metalworking, and electronics sectors, and more generally, sectors with substantial product heterogeneity, such as pharmaceuticals. User feedback is generally considered to be important in providing market knowledge, but it may provide technological knowledge as well. Technologically sophisticated and progressive users in some industries are capable not only of providing information on product performance and abstract user needs but also of helping to translate these needs into technical specifications, and even solving some technical problems.

16. One characteristic of the commercial failures reported in project SAPPHO differentiating them from innovations which succeeded was a relative lack of understanding of user needs. In general, this may lead to the absence of important product features or to the existence of unwanted features which may reduce overall performance (for example, excessive sophistication or even novelty). Some innovations covered by Teubal, Arnon, and Trachtenberg 1976 failed for these reasons. Rothwell's work also reports cases of failure due to lack of "manufacturability" of newly designed products (see Rothwell 1976a, 1976b).

17. There is also some evidence that manufacturing skills also make a direct contribution to design capabilities; for example, via knowledge of materials, testing procedures, and techniques associated with quality control. The introduction of preventive maintenance procedures to ensure better operation of an Argentinian steel plant, for instance, enhanced the firm's capabilities to modify and adapt machinery.

## Spin-offs versus explicit resource allocation

It is customary to distinguish between intangibles which are by-products from another activity and intangibles generated by a purposeful allocation of resources. The typical example of the former is production experience; that of the latter is technical knowledge acquired through R & D. It may be that this distinction is less useful for understanding the development of high-technology firms than one based on identifying sequences of activities or projects and the links (spin-offs) between them. Thus, the contribution of "doing" activity A to activity B may include intangibles accumulated both from doing and from explicit resource allocation, provided that they were specific to project A. Elsewhere the relevant earlier activity or projects have been identified as product innovation projects, and the contribution was termed a spin-off (Teubal 1982). A closer look at that case, however, and at other firms, shows the appropriateness of considering other explicit investments in R & D and in quality-control procedures (or explicit organizational changes) which, while not specific to a particular innovation, will have an effect on the profitability of later innovations. For example, establishment costs of an R & D laboratory or of an export marketing infrastructure would fall in this category. In one electronics firm, the costs of setting up the technical and other infrastructure for a new product class (II) explain the reduction in direct profitability observed in the short-run when shifting to that class.[18] This cost includes the need for mastering new technologies required for the newer and more complex product class, and physical investments. Therefore, the cost should not be attributed to the first generation alone.[19] Under the circumstances the profits of subsequent generations involve both a spin-off from past generations and a component attributable to past investments not specific to a product.[20]

## The economic value of intangibles

Very few attempts at measuring the value of specific intangibles accumulated by firms have been made, because of conceptual and data problems. In some

18. The actual shift is between the second generation of product class III and that of product class II, the latter being the first class II innovation that reached the market.

19. Considerations of this type have always been made with respect to investments in physical capital.

20. Strictly speaking, the indirect contribution of the earlier project to the later one—after having subtracted those expenditures which are clearly common to both—should not in its entirety be considered a spin-off. For example, operating profits of the earlier generation may have been low because the firm reduced prices to push sales in order to acquire reputation and feedback which would bring it large profits in subsequent generations. Part of the higher profits from the subsequent generations were thus planned and not obtained from nothing, so they should not be termed spin-off. Whether the indirect profitability or contribution of an activity is or is not a spin-off depends not only on the nature of the activity but also on whether the firm was looking only at short-run profits or at both short-run and long-run profits.

studies, such as those of Hollander (1965) and Enos (1962), a great deal of information has been collected and this could be used, in principle, to calculate rates of return. It seems that part of future effort at the microlevel should be directed toward this objective. The purpose here is to report on some preliminary calculations made for the electronics firm referred to earlier and to provide some relevant, though scant, information from two Brazilian capital-goods firms. Some of the factors which tend to make the value of intangibles high or low will then be indicated.

Table 7.2 reproduces the calculations performed for the electronics firm. The indirect profitability of the first generation of product III was assumed— on the basis of qualitative evidence—in its entirety to have benefited the second generation of III. An essentially similar procedure was carried out for product II. It was also assumed that the later projects would have been undertaken in the absence of the earlier projects, and that their discounted profits would have been zero (which in this case is an overestimate). The indirect profitability computed as a share of direct plus indirect profitability varied between 47% and 60% for product III, and between 29% and 48% for product II, depending on the assumptions made concerning markups and the share of R & D in total fixed project costs. These calculations are only of illustrative value, but they are consistent with the view of managers that interproject spin-offs were extremely important in the development of the firm. The numbers obtained are

Table 7.2. Direct and indirect profitability (electronics firm)

|  | Assumptions[a] | |
|---|---|---|
|  | (1) | (2) |
| *Profitability of product III first generation* | | |
| 1. Direct | 0.16 | 0.25 |
| 2. Total: direct *plus* second generation profitability attributable to first generation | 0.30 | 0.63 |
| *Profitability of product II-related projects*[b] | | |
| 3. Direct | 0.17 | 0.24 |
| 4. Total: direct *plus* fourth generation profitability attributable to prior projects[b] | 0.24 | 0.46 |
| *Indirect profitability as percentage of total* | | |
| Product III, (line 2 − line 1)/line 2 | 46 | 60 |
| Product II, (line 4 − line 3)/line | | |
|  | 29 | 48 |

[a]Let $m$ be average markup of unit variable costs and $\alpha$ the ratio of R & D to total fixed costs for the project. The value of $m$ is arbitrarily set in column (1) at 0.5 and in column (2) at 1. The ratio $\alpha$ is assumed to be 0.5 in both columns.
[b]Comprises the second and third product II projects and project IV. The base year was arbitrarily set at the base year of the latest of the projects in the set, the third product II project.

underestimates, although the contribution of early innovations to later ones has not been separated from the contribution of early nonspecific investments in R & D and infrastructure. The data collected from the two Brazilian firms interviewed and the data reported in other studies do not permit ready calculation of the value of accumulated intangibles. A major Brazilian producer of sugar-processing and alcohol-production equipment calculates that its past activity in the field is expected to save roughly $22 million in royalty payments, but that there are also other benefits derived from production experience. Another Brazilian firm, a successful pipe producer, stated that the value of the firm's experience was very large, but that it cannot be estimated from the company's accounts. It derives from shorter lead times in establishing new plants and in training the labor force, and from savings in royalty and other payments for technological transfer.[21] Finally the value of some of the investment and operating experience of the Usiminas steel plant of Brazil and of an Acindar steel plant in Argentina is probably very high since the significant capacity stretching undertaken has had a direct impact on the profitability of subsequent investments.

A number of factors explain why the value of spin-offs can be high. In our electronics firm, exogenous changes in the environment (especially the availability of new technology) coupled with a capacity to adapt to these changes have produced the effect mentioned. It should be noted that this case was characterized by a high rate of product obsolescence and also by a high rate of knowledge complementarity between old and new product classes. Thus, if a firm is willing to take the necessary risks by investing substantial resources in R & D and in marketing, it seems to stand a chance of maintaining or increasing its performance. Unwillingness to commit substantial resources to new projects exploiting newly available technology is probably a major reason for a low realized value of intangibles, especially in areas where prior knowledge, experience, and capability may be extremely important. In other cases, the science, technology, and experience underlying the new products appearing in the market may be completely different from those underlying current products. In these circumstances, the potential value of the accumulated intangibles is low, and the firm may opt to get out of the market. Finally, the value of intangibles may be affected by unexpected reductions in the size of the market and by government policies. For example, a design team embodying the firm's pool of design experience may have to be dissolved because of unexpected reductions in orders. These may result from macroeconomic policies such as reduced government procurement or currency overvaluation.

21. The firm also stated that its experience gave it enhanced planning capabilities which a newcomer would not possess.

Are patterns of intangibles accumulation industry-specific?

The patterns of intangibles accumulation presented above suggest that they are industry- or branch-specific; that is, a pattern based on investment experience was associated with the process and steel industries, one based on production experience was associated with metalworking, and one based on technical and market knowledge was linked with instruments and electronics. These patterns should not be overemphasized; they are probably more relevant for young firms than for big corporations, and to some extent, to firms in developing, rather than developed, countries.[22]

Thus, production experience is very important for electronic components and may be very important for producing certain kinds of electronic products, such as flat TV sets, where assembly is not the only or the main step. Similarly, the predominant accumulation of manufacturing skills in metalworking firms is probably more typical of firms active in protected markets and of young firms in developing countries. It is conceivable, however, that even in metalworking one can find young firms involved first in design and then in production (a pattern commonly found in electronics), especially if the original founders acquired production experience from employment in other firms.[23] Finally, in metalworking industries, intangibles associated with R & D and design may be of extreme importance, even in sectors which have traditionally been considered to be conservative, such as textile machinery. Rothwell (1976b) has shown that noninnovative firms in this area tended to fail despite their excellent experience and capability in manufacturing. The increased application of new chemical, electronic, and aerodynamic technologies resulted in a series of radical innovations which revolutionized the products of the industry. Thus, production experience, while still important and even a necessary condition for success, was no longer sufficient. New types of technological and market knowledge were becoming relevant, some of which required explicit R & D and the establishment of an R & D capability. This may also be the case in other metalworking and machinery branches.

Implications for innovation performance

The learning perspective to the development of high-technology firms, with its focus on the links and spin-offs between early activities or projects and later

22. This reflects the fact that most cases of firm learning processes seem to have been undertaken in developing countries.

23. The story of the Belgian firm Picanol seems to conform to this pattern. (See Rothwell 1976a.) In general, firms in developed countries may benefit from past experience of the *industry* (externalities), and therefore the learning in manufacturing need not be critical for every new metalworking firm.

activities or projects, may also be useful for analyzing innovation performance at the project level. The approach usually followed in analyzing innovation performance is to identify factors differentiating successful from failed innovations. For example, failed innovations may have resulted from insufficient investment in R & D, lack of interaction with customers, or management's failure to take into account other critical factors for success. The projects and most of the variables are usually not dated. The learning perspective would state that, in some cases, management may have attempted to act in the appropriate direction and may even have invested considerable resources. However, as a result of insufficient knowledge and experience (intangibles) accumulated from the past, the efforts of the firm were inefficient. The focus is on project selection, on whether the firm's response to the new requirements or opportunities fits in with its past experience (and capabilities). On the other hand, the commercial success of a particular innovation could be as much the result of the specific tactics adopted as of thoughtful project selection making full use of the intangibles and capabilities accumulated from the past. Thus, the analysis of any particular innovation is embedded in a view of the firm's development through time.

### Conclusions

With some notable exceptions, case studies of firms do not directly analyze the accumulation of intangibles, although they do refer to the role of experience in very general terms. Only one attempt has been made to measure the value of experience. Firm studies focusing on intangibles accumulation should (1) identify the main activities (investment spurts, products or R & D projects) characterizing the development of the firm, (2) define and compute the direct profitability of such activities, and (3) specify the main types or mechanisms of intangibles accumulation and attempt to calculate their economic value. An important aspect of (3) is the sequence of capabilities—from copying to wholly in-house designing, from production experience to the accumulation of a design capability. In particular, how does production/operating experience increase a design capability? A number of mechanisms have been introduced: preventive maintenance activities, which require intimate knowledge of the machinery (supply and demand factors); experience with available raw materials, which helps to specify targets for machinery modifications and adaptations; and similarly, use of specific machinery items that indicate the desirability of designing (or copying) and producing such machines in the future. Moreover, quality-control procedures enhance knowledge of materials and testing procedures and therefore pave the way for more basic R & D connected with materials.

A learning perspective to the development of high-technology firms may also contribute to identifying the relevant variables in empirical productivity, profit, or export equations. Cumulative R & D cannot by itself represent the process of intangibles accumulations, even in electronics, where, to some extent, it is the "engine of growth." This, apparently, is particularly true for the R & D associated with failed innovations. Both in electronics and in other industries, performance is highly dependent on intangibles other than those flowing from R & D, such as market knowledge, experience from investing, firm reputation effects, etc. The selection of an appropriate variable or variables depends on the industry or set of industrial branches considered and on the number of observations.

Production experience is probably not well represented by cumulative past output since an important part of it is increased capability to produce more sophisticated products. The main variables which would change skills and capabilities in this area are the level of sophistication achieved and explicit investments in purchasing and absorbing new technology. In dynamic product areas, such as electronics, it is extremely important to consider explicitly the effects of market creation and competitive pressures caused by the newly available technology. The capacity of a firm to adapt to these changes depends on the pool of intangibles available from the past and on current investments in R & D, marketing, and infrastructure. These factors will determine the extent to which the product profile of the firm will match the optimum profile, where the latter depends on technology and on competition. The next task is to suggest formal representations for these intangibles.

### References

Castano, A., J. Katz, and F. Navajas. 1981. "Etapas Historicas y Conductas Technologicas en una Planta Argentina de Maquinas Herramienta." Programa BID/CEPAL/PNUD de Investigariones sobre Desarollo Cientifico y Technologico en America Latina Monografia de Trabajo no. 38. Buenos Aires.

Dahlman, C. J., and F. Valadares Fonseca. 1978. "From Technological Dependence to Technological Development: The Case of the Usiminas Steel Plant in Brazil." ECLA/IDB/IDRC/UNDP Research Programme on Scientific and Technological Development in Latin America, Working Paper no. 21. Buenos Aires.

Enos, J. 1962. "Invention and Innovation in the Petroleum Industry." In *The Rate and Direction of Inventive Activity: Economic and Social Factors*, edited by R. Nelson. Princeton, N.J.: University of Princeton Press.

Griliches, Z. 1980. "Returns to Research and Development Expenditures in the Private Sector." In *New Developments in Productivity Measurement and Analysis*, edited by J. Kendrick and B. Vaccara. Chicago: University of Chicago Press.

Hollander, S. 1965. *The Sources of Efficiency Growth: A Case Study of the Du Pont Rayon Plants.* Cambridge: MIT Press.

Katz, J., et al. 1978. "Productivity, Technology, and Domestic Efforts in Research and

Development: The Growth Path of a Rayon Plant." Res. Prog. Sci. Tech. Dev. Latin America, Working Paper no. 13. Buenos Aires.

Maxwell, P. 1977. "Learning and Technical Change in the Steel Plant of Acindar, S.A. in Rosario, Argentina." Res. Prog. Sci. Tech. Dev. Latin America, Working Paper no. 4. Buenos Aires.

Rosenberg, N. 1976. *Perspectives on Technology*. Cambridge: Cambridge University Press. Chap. 1.

Rothwell, R. 1976a. "Picanol Weefautomation: A Case Study of a Successful Textile Machinery Builder." *Textile Institute and Industry* 14:103-6.

Rothwell, R. 1976b. *Innovation in Textile Machinery: Some Significant Factors in Success and Failure*. Occasional Paper Series, no. 2. Science Policy Research Unit, University of Sussex.

Rothwell, R., C. Freeman, A. Horsley, V. T. P. Jervis, A. B. Robertson, and J. Townsend. 1974. "SAPPHO Updated—Project SAPPHO Phase II." *Research Policy* 3:258-91.

Schmookler, J. 1966. *Invention and Economic Growth*. Cambridge: Harvard University Press.

Science Policy Research Unit, University of Sussex. 1972. *Success and Failure in Industrial Innovation: Report on Project SAPPHO*. London: Centre for the Study of Industrial Innovation.

Teubal, M. 1982. "The R & D Performance through Time of Young, High-Technology Firms: Methodology and an Illustration." *Research Policy* 11:333-46. (Chap. 6 in this volume.)

# 3   Government Policy, Externalities, and Growth

# Introduction to Part 3

Chapter 8 discusses the emergence and development of Israel's civilian high-technology sector, following the 1967 war. The basic mechanism used was the establishment of an industrial R & D fund administered by the Ministry of Commerce and Industry, which was thus independent of the traditional mechanism of support for science and higher education. The main characteristics of the promotion scheme were massive, *neutral*, and consistent support, at least during the first decade of growth. In particular, a 50% R & D subsidy was granted *any* project, i.e., belonging to *any* industrial branch, product class, or technological area, which passed the "minimum" criteria of technical feasibility (at first, the main criteria used) and a checklist of market factors. This scheme enabled a simultaneous process of *natural* selection of firms *and* "areas" to take place, one which is confirmed by the results of the biomedical electronics instrumentation study (chap. 1), where four out of the eight firms involved either collapsed, were reorganized, or abandoned the field. A number of additional conditions are mentioned, which contributed to what appears to be an efficient selection process that took place during the first decade of implementation of the system (when the above-mentioned neutral subsidy prevailed). The experience may be of value to semiindustrialized countries attempting to introduce promotional schemes for high-technology industry. One of its main attractions lies in showing how development may be consistent with providing incentives to many small entrepreneurs without "picking winners" in certain areas such as steel or other "basic" industries.

Chapter 9 proposes an overall normative framework for innovation policy in an open economy. Its starting point is the "market failure" set of justifications for government support of R & D, i.e., those based on the existence of significant indivisibilities, externalities, and uncertainties in the innovation process (see Arrow 1961; Nelson 1959). An attempt is made to pursue the analysis further by relating it to concrete policy, some examples of which were taken from case study work appearing elsewhere in this volume. In this respect, a fundamental distinction from the point of view of innovation policy is between *current* and *strategic* decisions. In our context strategic decisions relate first and foremost to investments in the scientific and technological infrastructure of the economy, that which directly stimulates and supports industrial innovation. These investments are indivisible; their planning and execution require a great deal of coordination within industry and between industry and government (since the availability of these investments generates considerable user surpluses); and there is a simultaneity problem between them and what Hirschman (1958) would call directly productive activities. Current decisions, on the other hand, are not characterized by strong indivisibilities and refer to industrial R & D project support such as that being carried out in Israel and described in chapter 8. While the principle of neutrality in government support (no preference for product class, technological area, or industrial branch) is indeed relevant and applicable to a large extent in relation to current decisions, it is not relevant, even in principle, in relation to strategic decisions (given the inherent indivisibilities). Therefore, in some sense, there is less room here for a natural selection mechanism to operate. It may be necessary to make hard choices in relation to the types of science and technology infrastructure a country should invest in, particularly a small country—e.g., should the emphasis be in information technology, in new materials, or in biotechnology?

Chapters 10 and 11 relate to what in innovation policy terms are strategic decisions. United States government support of communications satellites, which led to the 1965 launch of the first geostationary communications satellite, should be regarded as support for a particular type of science and technology infrastructure (one in which the basic launch and transmission capabilities were proven). The historical description of NASA's role in this extremely successful case of government support is one example of a model of decision making where a very important role is assigned to the private sector. One of the effects of government's support was an acceleration of the availability and commercial introduction of communications satellite technology (at least by five years). Chapter 11 presents a formal normative framework for discussing the optimal private and optimal social times for the introduction of a new technology into the economy (given its availability). A gap between the two exists because of the externalities generated by technology introduction (in sector 1) that facilitate its subsequent diffusion (into sector 2). The "marginal externality gain" from accelerating the introduction of the technology is the basis for the technology introduction subsidy allowed for sector 1. The paper also discusses the economic benefits from accelerating the "availability" of the new technology ("availability" would in the satellite communications case, represent the science and technology infrastructure generated by NASA prior to commercialization). It suggests that these benefits may be substantial, especially when the efficiency of conventional technology in the introducing sector (e.g., cable technology as opposed to satellites) is increasing through time. In some cases, such an acceleration is critical for the eventual introduction and diffusion of the technology

In all of our discussions up to now on innovation policy issues, both current and strategic, there has been almost no link with the wider issues of industrial policy generally and economic growth. This is particularly bothersome since one objective of innovation policy is the stimulation of economic growth. (for a discussion of a number of issues relating innovation to economic growth, see Nelson 1984). In order to achieve such an integration a broader framework relating economic growth, structural change, and innovation (broadly speaking, including development of the science and technology infrastructure) is required. Without such a "realistic" growth model of the economy it is impossible to assess the quantitative framework for innovation policy. Modern growth theory, with its emphasis on capital accumulation, is incapable of dealing with the central issues involved. What is needed is a truly dynamic theory of growth, which sets the economy in an evolving global context in relation to trade, markets, technology, and competition, and which focuses on structural change and on dynamic comparative advantage. Such a model, in my opinion, should emphasize, among other features, the pervasiveness of externalities; the central role of certain technologies (e.g., information technology in this decade); the critical role of certain inputs for growth (such as a plentiful supply of qualified scientists, engineers, and technicians); and the dynamics of infrastructure development. Judicious government intervention in relation to the above innovation-policy-related features should be combined with a strong export and outward orientation of the economy and with maximum exploitation of market forces, dynamic comparative advantages, and entrepreneurship. A number of independent developments are gradually setting the stage for such a breakthrough. Chapter 12 presents a very rudimentary and simplistic growth model, one where a pivotal role is played by the so-called engineering (capital goods) sector. Its interest derives, I believe, from a number of unconventional features, although in its present form it bears almost no resemblance to the model of growth suggested above. By means of an analysis of the growth path of an export-led staple economy, it addresses two main issues: the conditions (and time) of emergence of the engineering sector; and the conditions under which such a sector will perform a *key-sector* role, i.e., that of adapting technologies and diffusing techniques across other sectors (see Rosenberg 1961). In our context, this means the adaptation of techniques available from the world's supply of techniques to the requirements of local production, thereby enabling the economy to establish new productive activities. Much still remains to be done in order to begin to address the main issues of innovation policy within a dynamic growth context.

## References

Arrow, K. 1961. "Economic Welfare and the Allocation of Resources for Invention," In *The Rate and Direction of Inventive Activity*, edited by R. Nelson, 609–25. Princeton: Princeton University Press.

Hirschman, A. 1958. *The Strategy of Economic Development*. New Haven: Yale University Press.

Nelson, R. 1959. "The Simple Economics of Basic Scientific Research," *Journal of Political Economy* 67:297–306.

Nelson, R. 1984. "Policies in Support of High Technology Industries," Working Paper no. 1011 (revised), Institution for Social Policy Studies, Yale University.

Rosenberg, N. 1963. "Technological Change in the U.S Machine-Tool Industry: 1840–1910." *Journal of Economic History* 23 (December); 414–43. Reprinted in *Perspectives on Technology*, chap. 1. London and New York: Cambridge University Press, 1976.

# 8 Neutrality in Science Policy
## The Promotion of Sophisticated Industrial Technology in Israel

## Morris Teubal

The current system of promotion of industrial research and development in Israel began in 1967 with the establishment of the Industrial Research Fund in the Ministry of Commerce and Industry. The objective of this fund was to subsidize civilian research and development performed in the industrial sector, without any explicit preferences for any particular industrial branch, technological area, or class of product. A flat subsidy of 50% of all expenditures on research and development by the firm making the application was granted to all projects submitted to the ministry and fulfilling a set of minimum requirements. These originally involved proof of technical feasibility, and account was taken of the reputation of the scientists and engineers involved; only later on and gradually was information on the state of the market and the likely competitors required for approval. At present, there are requirements of a minimum of satisfactory answers to 22 questions in the applications to the ministry for grants. These questions deal with the marketing and other capacities of the firm and with its marketing plan.[1]

The original minimal requirements did not explicitly affect approval of grants in favor of any particular branch, product, or technology, at least during the first decade of existence of the system. Thus, the system of support for industrial technology was at least formally neutral in this respect during that period.

### General Evolution of the System
### of Research and Development in Israel

The emphasis on science and research is deeply rooted in the history of Israel (Arnon 1978; Dudai 1974; Hershkovitz 1980), since the arrival of the first immigrants from Eastern Europe in the last quarter of the nineteenth century.

1. Dr. A. Lavie, chief scientist of the Ministry of Commerce and Industry at a conference on science policy, June 1981.

At first, the emphasis was on applied research oriented towards solving concrete problems confronting the new society, e.g., epidemiological research and research in agriculture; the first agricultural experimental research station was established in the early years of the twentieth century. A more varied picture emerges from the period of the Mandate, from 1918 to 1948, when the main institutions of scientific research were established: these were the Hebrew University of Jerusalem, the Technical University of Haifa (The Technion), the Agricultural Research Station of Rehovot, and the Ziv Institute of Science, which later became the Weizman Institute. Scientific activities during the period of the Mandate were mainly concerned with the study of characteristics of the land—climate, soil, water resources—plant and animal life, plagues and illnesses, geography and geology. Basic research, conducted in accordance with the Western tradition of academic freedom, and agricultural research achieved a high level relative to the standards of the period.

There was practically no industrial research, except for some connected with the Dead Sea Works. After Independence, Prime Minister D. Ben-Gurion himself headed the Research Council founded in 1949; its objective was to extend further the institutional structure of scientific work in Israel. The new government established a number of governmental research laboratories during the 1950s, for example, the Fibers Institute, the aim of which was to support the textile industry being developed to provide employment to the more than half a million Jewish refugees from Arab countries, and the National Physics Laboratory. It also established several new universities—Tel-Aviv, Bar Ilan, Negev—and founded other institutions such as the National Council for Research and Development (NCRD) in 1959 and the Israel Academy of Sciences in 1961.

By the mid-1960s, institutional arrangements for scientific work were well under way; additional efforts were made to reinforce existing institutions rather than to establish new ones. This period shows also the first attempts at a more systematic approach towards research and development. Thus among the objectives of the National Council for Research and Development we find planning governmental policy towards research and development and defining "national research needs" in various fields. The most significant event, however, was the nomination in 1966 of a committee for the organization and administration of government research—the Kachalsky committee. The committee's recommendations were: that bureaus of chief scientists be created in ministries such as the Ministry of Commerce and Industry in order to coordinate their activities in research and technology and to stimulate applied research; that the governmental research institutes be reorganized into three research authorities, each headed by the chief scientist of the corresponding ministry, e.g., the Ministries of Agriculture, Commerce, and Industry, and

Development; that the National Council for Research and Development be organized in a way which would enable it to perform such functions as the design of a national policy for research and development, to deal with scientific manpower, and to coordinate the activities of the various chief scientists.

The first of these recommendations was the most significant since it led to substantial increases in applied research and development, especially in industry by means of the Industrial Research Fund of the Ministry of Commerce and Industry. The 50% rate of subsidy of all approved projects in research and development was not part of any law or regulation governing the functioning of this fund, which was probably viewed as a scheme that was simple to administer and involved a reasonable distribution of the resources for research and development between governmental and industrial firms. The Kachalsky committee's last recommendation was not carried out; each ministry in fact acted independently, with the total government budget for research and development being simply the total of the budgets for research and development of each ministry—each being part of the total budget which the individual ministries negotiated with the Treasury. The magnitude and structure of civilian research and development do not seem to be the result of an explicit policy framed in quantitative terms.

### The Promotion of Industrial Research

Until the mid-1960s, almost no civilian research and development was conducted in private industry in Israel, and there was no governmental support of such research as there was (Dudai 1974). Most of the existing industrial research was connected with the exploitation of deposits of potash and bromine in the Dead Sea.

Two main developments probably account for the emergence of new arrangements for governmental support in 1967. The major institutions and arrangements linking scientific research and technology were already well established. Israeli scientists and engineers were engaged in research in universities, in agriculture, and in defense, but around that time it began to become clear that they could also be employed in research and development in private industry producing goods for civilian markets. The other factor which led to the new pattern emerging after 1967 was the realization that the attainment of Israel's objective of increasing its income from exports required a major shift in the distribution of economic activities. The continued growth of exports could not be achieved by concentration on the exportation of oranges and textiles; existing exports of these products were being threatened by increased competition coming from producers with lower costs. It was necessary to increase the technological or scientific content of products through an increase in the intensity of research and development, design and marketing; this was thought to be the only way to overcome the competition based on lower costs. The alternative was to increase the size of plants in traditional industries, but this would have involved too much risk, given the small size of the domestic markets.

Since sophisticated exports are not an objective in themselves, governmental intervention to promote the underlying industrial technology can be justified only in terms of some kind of "market failure" in the production of such goods. Technologically sophisticated exports require the development of a set of investments in at least a minimum of organizations and manpower for research and development. Governmental support can easily be justified in terms of the unpriced benefits which any private firm undertaking such investments would provide—positive "externalities"—to other firms or users. Another reason for governmental support of this investment in organization and staff for research results from the fact that capital markets were imperfect in the period considered. Japanese policies directed to similar ends included the subsidy of costs of penetrating certain export markets (Magaziner and Hout 1980).

The first chief scientist was a university professor with a background in medicine and research. His work included research on cancer. The second chief scientist was a former army officer with a degree in engineering who has previously headed the department of research and development of the Ministry of Defense. His experience in linking research and development performed in the defense sector to the needs of the armed forces was probably responsible for the increased emphasis given to the market and to marketing by the Office of the Chief Scientist in approving projects. Another factor was the accumulated collective experience showing the critical role of these facts in the commercial success of technological innovations. The main features of the system of promotion of industrial technology, which affected its subsequent performance, were its adherence to a policy of neutrality with respect to branch of industry, technological field, or class of product; and its concentration on the support of civilian research and development directly performed in industrial firms. Its location in the Ministry of Commerce and Industry made it an integral part of general industrial policy.

## Review of Performance of Research and Development in Civilian Industry

The main trend in civilian research and development in the natural sciences and engineering in the first years of the new policy was a very significant increase in total expenditure on civilian research and development from $34 million in 1965 to over $230 million in 1978. There was an almost threefold increase in the share allocated to civilian research and development performed in private industry out of total expenditure on such activities. This share rose

Table 8.1. Civilian research and development expenditure on natural sciences and engineering, including agriculture, mathematics, and medicine

| Year | Total (current U.S. $ in millions) | Distribution of all expenditure among institutions conductiong R & D [a] | | | Governmental share in expenditures | | |
|---|---|---|---|---|---|---|---|
| | | Industrial firms (%) | Academic institutions (%) | Government (%) | Total R & D [b] (%) | R & D in industrial firms (%) | R & D in academic institutions (%) |
| 1965–66 | 34 | — | — | — | — | — | — |
| 1966 | 42 | 11 | 62 | 27 | 51 | 16 | 39 |
| 1967 | — | — | — | — | — | — | — |
| 1968 | — | — | — | — | — | — | — |
| 1969 | — | — | — | — | — | — | — |
| 1970 | 70 | 21 | 62 | 17 | 57 | 25 | 56 |
| 1971 | — | — | — | — | — | — | — |
| 1972 | — | — | — | — | — | — | — |
| 1973 | 120 | 19 | 62 | 19 | 59 | 24 | 59 |
| 1974 | 127 | 24 | 62 | 14 | 62 | 29 | 66 |
| 1975 | 115 | 23 | 59 | 18 | 62 | 32 | 65 |
| 1976 | 183 | 35 | 53 | 13 | 56 | (61)[c] | — |
| 1977 | 220[d] | — | — | — | — | — | — |
| 1978 | 230[d] | 43 | 45 | 12 | — | (56)[c] | — |

*Sources: Science and Technology in Israel 1975–76* (Jerusalem: National Council for Research and Development, 1977); Hershkovitz 1980.
[a] Agricultural research and development is performed in governmental laboratories.
[b] Includes civilian research and development in the social sciences.
[c] Percentage of governmental funds in total research and development (including military) performed in industry.
[d] Lower boundaries, assuming share of governmental funds to be 50%.

from 11% in 1966 to 43% in 1978 (table 8.1). Expenditure on civilian industrial research and development rose from $12 million in 1969 to $75 million in 1977 (in current dollars). The number of qualified scientists and engineers in research and development rose from 886 in 1969 to over 3,000 in 1981 (table 8.2).

The growth in civilian research and development carried out in industry is partly accounted for by the growth of support by the government. The share of governmental expenditure in the total expenditure on civilian industrial research and development rose from 16% in 1966, before the establishment of the new system, to 32% in 1975; other figures show an increase from 12% in 1971 to nearly 50% in 1979. The grants of the Ministry of Commerce and Industry rose from $1.2 million to $32 million in 1979 (table 8.3). In 1969 there were 210 industrial establishments performing civilian research and development (table 8.4). The numbers fluctuated considerably during the next

Table 8.2. Scientists, engineers, and technicians in civilian industrial
research and development[a]

| Year | Qualified scientists and engineers | | Practical engineers and technicians | | Total skilled manpower in R & D |
|---|---|---|---|---|---|
| | Number | Percentage of all persons employed in industry | Number | Percentage of all persons employed in industry | |
| 1966 | — | — | — | — | — |
| 1967 | — | — | — | — | — |
| 1968 | — | — | — | — | — |
| 1969 | 886 | 0.45 | 671 | 0.34 | 1,557 |
| 1970 | 1,013 | 0.49 | 999 | 0.48 | 2,012 |
| 1971 | 1,141 | 0.51 | 1,124 | 0.51 | 2,265 |
| 1972 | 1,254 | 0.53 | 1,259 | 0.54 | 2,513 |
| 1973 | — | — | — | — | — |
| 1974 | 1,438 | — | 1,105 | — | 2,543 |
| 1975 | 1,653 | 0.66 | 1,410 | 0.56 | 3,063 |
| 1976 | 2,052 | 0.79 | 1,649 | 0.64 | 3,701 |
| 1977 | 2,212 | 0.84 | 1,669 | 0.64 | 3,881 |
| 1978[b] | (1,013) | — | (987) | — | (2,000) |
| 1979[c] | 2,600 | — | 3,200[c] | — | — |
| 1980 | — | — | — | — | — |
| 1981 | 3,000[d] | — | — | — | — |
| 1982 | — | — | — | — | — |

*Sources:*
[a]Central Bureau of Statistics, *Survey of R & D in Industry 1977–78*, Research and Development Statistics Series 13, reprinted from Supplement to *Monthly Bulletin of Statistics* 11 (1980).
[b]Figures for 1978 are full-time equivalents. Data from Office of the Chief Scientist, internal sources.
[c]From Ministry of Industry, Trade, and Commerce, *Industry R & D Opportunities of Israel* (Jerusalem, 1977).
[d]Private communication from National Council for Research and Development.

six years without any significant net growth, attaining a maximum of 228 in 1975. There was a significant growth in numbers after 1975, reaching about 500 or more industrial establishments performing research and development in 1980.

The results of this expansion of activity in research and development in private industry may be seen in the figures on exports from this sector. We are at present in no position to undertake a realistic analysis of costs and benefits of the introduction and development of the Israeli system of promotion of industrial technology. Exports from the industrial sectors utilizing sophisticated

Table 8.3. Growth in governmental support of research and development in industry

| Year | R & D grants to industry from Ministry of Commerce and Industry | Exports resulting from grants for R & D (current U.S. $ in millions) | Total exports from technologically sophisticated industries |
|---|---|---|---|
| 1967 | 1.2 | 1.6 | — |
| 1968 | 1.5 | 3.7 | — |
| 1969 | 2.5 | 5.9 | — |
| 1970 | 3.0 | 8.0 | 158 |
| 1971 | 3.5 | 10.2 | 196 |
| 1972 | 3.7 | 20.8 | 220 |
| 1973 | 5.4 | 100.9 | 266 |
| 1974 | 9.0 | 233.4 | 478 |
| 1975 | 10.0 | 289.9 | 530 |
| 1976 | 20.0 | 283.6 | 756 |
| 1977 | 25.2 | 416.3 | 985 |
| 1978 | 27.0 | 550.0 | 1,284 |
| 1979 | 32.0 | 750.0 | 1,680 |
| 1980 | — | — | 2,143 |
| 1981 | — | 1,000.0 | — |
| 1982 | 60.0 | 1,400.0 | — |

*Sources:* Office of Chief Scientist, Ministry of Commerce and Industry, *Industrial R & D* (Jerusalem, 1976); N. Guttentag, "The Effect of R & D on the Structure of Industry: Part I: R & D in Israeli Industry—Inputs and Outputs" (typescript) (December 1981); Central Bureau of Statistics, *Statistical Abstract of Israel* (Jerusalem, various years); Bank of Israel, *Annual Reports* (Jerusalem, various years).

and intensive research and development, such as electronics, transportation equipment, chemicals, metal products, machinery, and rubber and plastics, which have received most of the subsidies for research and development, have increased very markedly. Exports from projects supported by the Ministry of Commerce and Industry grew from $1.6 million in 1967 to $750 million in 1979 and to over a billion dollars a short while later. This figure understates Israeli exports from industries of high technology because there are exports from these sophisticated industries sectors which are not the result of research and development or of research and development projects supported by the ministry. There can be no doubt that the composition both of industrial exports and of output changed considerably in the 1970s. Between 1970 and 1980, industrial exports, excluding diamonds, grew at an average real rate of 11%; nominal industrial exports, except diamonds, rose from nearly $400 million in 1970 to over three billion dollars in 1980. In the same period, the share of the technologically sophisticated sectors increased from 40% to 66%

Table 8.4. Research and development activity in private industry

| Year | Industrial[a] establishments conducting R & D (number) | Continuing[b] projects (number) | Industrial contracts with universities (current I £ in millions) |
|---|---|---|---|
| 1969 | 210 | — | — |
| 1970 | 308 | — | — |
| 1971 | 273 | — | — |
| 1972 | 294 | — | — |
| 1973 | — | 200 | 2 |
| 1974 | 216 | — | 5 |
| 1975 | 228 | 400 | 10 |
| 1976 | 289 | — | 20 |
| 1977 | 305 | — | — |
| 1978 | — | 581 | — |
| 1979 | 350[c] | — | — |
| 1980 | 500[d] | 600 | — |
| 1981 | 300–400[e] | 1,000[f] | — |

*Sources:*

[a]Figures until 1977 are from Central Bureau of Statistics, *Survey of R & D in Industry 1977–78.*

[b]Various internal sources, Ministry of Commerce and Industry. They refer to projects supported by the ministry only.

[c]From National Council for Research and Development, *Science in Israel* (Jerusalem, 1979).

[d]From Ministry of Commerce and Industry, *Industrial R & D Opportunities in Israel (Jerusalem, 1980), 3.*

[e]Companies receiving support from the Office of the Chief Scientist (presentation of Dr. A. Lavie, June 1983).

[f]Estimated number, communication from the Office of the Chief Scientist.

(table 8.5). The average annual real growth of exports of sophisticated sectors was 17% during the 1970s, while it was only 6% for exports of the conventional industrial sectors.

In 1970 only one technologically sophisticated branch of industry, namely, chemicals, was in the first five exporting industries. In 1979 there were three such industries: chemicals, transport equipment, and metal products. The rank of the electronics industry rose from ninth to sixth place. (The data on the electrical and electronics branch underestimate the contribution of the all-pervasive electronics technology.) Textiles fell from fourth to eighth place in the 1970s. The rate of growth of output and of the share of output exported was higher during the 1970s in the technologically sophisticated industries than in the more traditional ones. The average share of output exported by the former at the end of the decade of the 1970s was about 45%, while that of the traditional exporting industries was only 26%, reversing their ranking of 1970

Table 8.5. Basic features of sophisticated (*S*) and traditional (*T*) industries

| Year | Total exports (current U.S. $ in millions) | | | Share of *S* in exports (%) | Share of *S* in total output (%) | Share of output exported (%) | | |
|------|------|------|------|------|------|------|------|------|
| | *S* | *T* | Total | | | *S* | *T* | Total |
| 1970 | 158 | 236 | 394 | 0.40 | 0.43 | 16.6 | 21.1 | 19.2 |
| 1971 | 196 | 281 | 477 | 0.41 | 0.45 | 17.7 | 23.8 | 21.0 |
| 1972 | 220 | 307 | 527 | 0.42 | 0.45 | 16.2 | 22.2 | 19.5 |
| 1973 | 266 | 360 | 626 | 0.42 | 0.45 | 17.2 | 20.4 | 19.0 |
| 1974 | 478 | 454 | 932 | 0.51 | 0.47 | 21.0 | 20.8 | 20.9 |
| 1975 | 530 | 440 | 970 | 0.55 | 0.48 | 20.3 | 18.9 | 19.6 |
| 1976 | 756 | 472 | 1,228 | 0.62 | 0.49 | 28.6 | 19.5 | 23.9 |
| 1977 | 985 | 965 | 1,553 | 0.63 | 0.49 | 35.1 | 21.3 | 28.0 |
| 1978 | 1,284 | 638 | 1,922 | 0.67 | 0.49 | 38.8 | 19.6 | 28.9 |
| 1979 | 1,680 | 818 | 2,498 | 0.67 | 0.48 | 39.7 | 21.2 | 30.2 |
| 1980 | 2,143 | 1,121 | 3,264 | 0.66 | 0.48 | 45.0 | 26.3 | 35.2 |

*Source:* Calculations from data published in Central Bureau of Statistics, *Statistical Abstract of Israel* (Jerusalem, various years).

(see table 8.5). The growth of output and exports in industries with intensive research and development since 1967 is sufficiently impressive to justify a presumption that the arrangements for the support of industrial research and development have been successful.

Before proceeding it is important to mention that the founding of a number of firms after 1967 and their success after this date were favorably stimulated by support received from the Office of the Chief Scientist. This was particularly true for electronics firms. Among such firms are Elbit—a minicomputer firm "spunoff" from the defense sector; Elscint—which initially specialized in nuclear instrumentation generally, then more specifically in nuclear medical technology, and, since 1977, diversified into computerized axial tomography and ultrasound medical instruments; AEL—a partly foreign-owned firm producing microwave components and communication systems; Scitex—a firm developing and producing computer-aided design systems for the textile, printing, and electronics industries; Beta Engineering—a firm initially producing specialized instruments for a wide variety of fields ranging from measurement of blood pressure and dental technology to numerically controlled sewing machines. In the chemical and plastics industry, there were 20 firms active in research and development in 1971. These were concentrated in pharmaceuticals, basic chemicals, pesticides, fine chemicals for laboratories, and detergents. Research and development was concentrated, however, in a few firms, such as Machteshim owned by Histadrut and founded in 1952.

## The Nondiscriminatory (Neutral) Character of the Incentives

The system of promotion, established in 1967, offered the same rate of subsidy for research and development for all approved projects, regardless of the branch of industry, class of product, and technological area. An approved project in textiles would receive proportionally the same grant for research and development as one in electronics. If there was a bias, it was in favor of projects for research and development that might lead to exports; these received a subsidy of 50% compared to projects that might lead to import substitution, which received a subsidy of only 25%. This formal discrimination was not, however, a real one since all projects had to be "directed" towards exports, as was explicitly stated in the application forms for grants for research and development. Discussions with officials responsible for deciding about the applications for grants in the Office of the Chief Scientist and an analysis of the share of total grants in the total research and development budgets of projects approved by the Office support this view of the neutrality of the policy. This share approached 50%, even when a significant number of the approved projects did not in fact lead to exports. Related to this point is the fact that the policy followed in the early years of the system was one of "force-feeding" of grants—the Office of the Chief Scientist was more interested in increasing the number of firms doing research and development, and in extending the kinds of research and development the firms engaged in, than in controlling the fulfillment of the specific obligations accepted by the firms, including that of exporting their products.

The policy of neutrality has changed; the change began in 1976. The first major change was the introduction of the National Programmes scheme, which provided for a higher rate of subsidy for research and development than that granted under the scheme then existing. While the new scheme did discriminate among firms—it was open only to firms who had succeeded in the past— it still was neutral with respect to branch. The projects approved had to be relatively large, to involve large risks, and to have a high expected return. Thus, there might have been some, probably justifiable, departures from strict neutrality with respect to class of product or technological field. Most of the funds of the new scheme have apparently gone to electronics, and some of the projects supported have been significant commercial successes.

The National Programmes were not intended to compensate for deficiencies in the work of the Office of the Chief Scientist and the policy of neutrality. They were a response to changed circumstances, especially the emergence of a group of firms which had succeeded in the past, and the potential contribution of which to the expansion of exports was estimated to be great, provided special support could be given for research. This scheme has since been discontinued: most of the firms which benefited from it now have access to the local and sometimes to the international capital markets. More significant departures from neutrality occurred in the late 1970s and early 1980s.

### Formal and Effective Neutrality

The formal, or nominal, neutrality of a uniform rate of subsidy for all projects of research and development is not necessarily equivalent to effective neutrality in the promotion of innovations. It certainly does not treat all sectors of industry equally; there are considerable differences in the proportion which research and development makes up in the total costs of innovation in the various industrial sectors, types of technology, or classes of product. A uniform rate of subsidy for research and development favors electronic innovations more than innovations in the chemical industry, since, in the latter, investments in plant are a larger part of the total costs of innovation (Mansfield et al. 1971, 110). It does not seem to favor electronics innovation more than innovation within the mechanical engineering industry. Effective neutrality with regard to innovation need not mean effective neutrality with regard to the different sectors of industry, since the proportion of the investment of research and development to sales or to total investments may vary from one sector of industry to another.

There are two additional reasons why formal neutrality does not necessarily result in effective neutrality in Israel. First, grants for research and development for the military are not neutral vis-à-vis civilian technology since they are presumably directed to particular kinds of technology, especially in electronics, and particular classes of products such as communications equipment. At most, we may say that the promotion of civilian technology is nominally neutral, if one sets aside the nonneutral development of military technology. Thus, while the high proportion of the approved projects in electronics did not result from a preference, on the part of the Office of the Chief Scientist, for civilian electronics over other civilian projects, it did in part result from the fact that the development of electronics technology was given special preference in the defense sector. Furthermore, the "minimum requirements" for approval of projects may have implications of nonneutrality, because the fulfillment of these requirements is much simpler in some industries than in others.

### Evidence of Neutrality

Has the system for the promotion of industrial technology really been neutral? The system in Israel was—at least formally—neutral during the first decade of its existence. This is evident not only from the regulations governing the activity of the Office of the Chief Scientist, but also from the accounts given by officials responsible for allocating the grants, and from some statistical data on the proportion which government provided for the support of research and development in a variety of sectors of industry. There was no shortage of funds at least until 1975; the total expenditures on grants for projects meeting the minimal requirements were always smaller than the funds available, even when the number of approved projects increased (table 8.3). In 1976 governmental support for research and development made up approximately similar proportions of the total expenditure on research and development in the various branches of industry. They range from 28% for chemicals and oil to 37% for basic metals and metal products. In electronics and scientific instruments, it was 35%, while in rubber and plastics it was 33% (Pakes and Lach 1982). There was wider variation between branches of industry in 1970, but this may be explained by obstacles incidental to the beginning of the system, such as the insufficient knowledge of new firms about the opportunities for obtaining grants.

There is clear evidence of effective nonneutrality in the policy for promoting innovation during the second half of the 1970s. After a certain point, a growing shortage of funds did not enable the scheme for grants of 50% for research and development to be followed completely, and a number of other criteria were introduced. Some of these changes in policy were probably intended not to be neutral; there was, for example, an apparently increasing preference for the support of research and development in electronics at the expense of other projects in other industries such as chemicals. thus, in 1979 while 32% of all research and development in electronics and scientific instruments was financed by the government, only 4% of research and development in rubber and plastics was so financed.

### Desirability of Departures from Neutrality

The Israeli system of promotion of industrial research began without any explicit preference for particular classes of products, branches of industry, or types of technology; it then gradually began to depart from that neutrality. It did so apparently in consequence of a financial constraint.

The departure from strict neutrality could be justified on grounds of principle; there were also grounds of expediency. At the beginning there was, at best, very little information on the prospects of success of particular projects involving research and development in any one firm for the profitability of other firms. Under these circumstances, a policy of neutrality seemed better than any alternative; furthermore it did not stifle initiative in the use of research. One of its important advantages in the early stage was that it permitted the accumulation of a wide variety of experience and information. Once acquired and assessed, this information and experience could indicate certain branches which had better prospects for commercial success for themselves and for beneficial effects for other firms. The policy of neutrality might then be revised.

My inclination is to justify only moderate departures from neutrality, except where there is a severe financial constraint. It is naïve to assume that enough information can be collected to justify a radical departure from the principle of neutrality. A complete abrogation of neutrality could stifle the emergence of initiative and creativity in the less-favored fields. The spirit of this conclusion is similar to that reached by L. Westphal, who in the area of industrial policy in general has emphasized the desirability of providing strong preferential support for a small number of industries rather than to a very wide range (Westphal 1982).

Although the first decade's neutral policies were probably justified, it does not follow that the departures from neutrality actually followed later on by the Office of the Chief Scientist were adequate or optimal. To the best of my knowledge, not enough effort was made to collect, organize, and analyze information about the experience of the early years. Thus, it might have been that the office was not in the best position to predict "winners" among the various industries or those where externalities might be particularly large.

### Strategies for the Promotion of Industrial Technology

There were two alternative ways in which the newly created system for the promotion of industrial technology could have been established. One was to incorporate it into existing research institutions like universities and research institutes, through the support of higher education and science; the other was to place it in the Ministry of Commerce and Industry.

Similarly, there are two possible direct beneficiaries of governmental efforts to promote industrial technology: the universities and governmental laboratories (research institutions) on the one side, and business firms on the other. The idea behind the first set of alternatives is that universities and governmental laboratories would develop prototypes which would then be transferred to industry. Industrial firms would then produce on a commercial scale prototypes developed elsewhere. Thus, a large share of governmental support for industrial research and development would go to the academic and independent research institutions for joint projects in which those institutions collaborated with industrial firms. This alternative might have given too much emphasis to the aspirations and demands of the research institutions rather than to the promotion of "high-technology" industries.[2] Thus the criterion of technological or scientific "novelty" or originality rather than the criterion of "commercial prospects" might have become paramount.[3]

The central feature of the Israeli system for the promotion of industrial

2. On the case of Norway, see Irvine et al. 1982.

3. In this connection, while technological creativity may contribute to commercial success, it would be erroneous to base the promotion of emerging systems of industrial technology wholly on criteria of scientific and technological originality. See Teubal, Arnon, and Trachtenberg 1976.

technology since about 1967 has been its direct support to business firms and its location within the Ministry of Commerce and Industry. It was a departure from hitherto existing arrangements for the support of research in scientific and higher educational institutions, and to a large extent it was able thereby to avoid the dangers of an excessively academic approach to research and development. The Office of the Chief Scientist reduced the share of funds for research and development in the semiautonomous industrially oriented governmental laboratories and shifted resources to support research and development performed by private industrial firms in their own laboratories. The share of total civilian research and development in natural science and engineering performed in governmental laboratories declined from 27% in 1966 to 12% in 1978 (table 8.1).

A system for the promotion of industrial technology on the lines followed in Israel—oriented to business firms and centered in the Ministry of Industry and Commerce rather than in a Ministry of Science—does not imply that the development of an industry using high technology and manufacturing high-technology products does not have certain scientific and academic preconditions. But how much and what kinds of investments in academic and governmental laboratories and expenditures for scientific training at home and abroad are required? Support of the development of the basic scientific and technological prerequisites—including manpower—should not be the main target of a governmental scheme for supporting the development of technology for use by Israeli industries, although such support is important and even critical in order to develop capacities in new fields of technology like robotics, bioengineering, and electronics.

The separation of the promotion of industrial technological research from the support of higher education and science in universities may in fact be in the best interests of the scientific community in the long run. An industry successfully using high technology will require increased numbers of scientists and engineers as well as research institutes and, by producing a demand for them, will contribute to their development. Special social, economic, and political factors led Israel to introduce a particular system for the promotion of industrial technology that differed from the type of system that emerged in a number of other countries. In the middle of the 1960s, the difference between scientific research and industrial technology and the understanding that the latter is not the automatic result of the former were already well established in the minds of the Israeli authorities. Despite Israel's distinguished achievements in scientific research, the recommendations of the government committee, headed by the eminent Israeli scientist Kachalsky, decided on the location of the system of promotion within the Ministry of Commerce and Industry. This decision was affected by reflection on what had been achieved in agriculture and defense, where, through considerable practical experience, scientific research had been effectively harnessed to the advancement of technology (Katz and Ben-David 1975).

### Implications of Neutrality

#### "Natural selection"

A consequence of the policy of neutrality was the initiation of a process of natural selection simultaneously both of firms and of industrial fields. The process of selection may be designated as "natural" because one determinant of a firm's survival was its competitive success in the market, and not its success in being pleasing to officials by agreeing to undertake projects of a particular type or in belonging to a particular industrial field which officials regarded as urgent. "Natural selection" is more efficient than other selective mechanisms when there is little information in advance about which areas will be "winners" or about which will have especially valuable spin-offs or externalities; this has been the case in the early stages of industries using and producing high technology. The process of selection in Israel occurs simultaneously among firms and among industrial fields. Israel could not have developed an advantage in computerized tomography without having an unusually capable private firm which decided to enter this area. It was reasonable to believe that Israel should develop an industry of high technology, but it was not possible to ascertain simply on the basis of the relative abundance of skilled manpower which particular branch of industrial high technology or which firms would be profitable. Much has depended on the capacities of the entrepreneurs entering the various branches. Nor is the survival of enterprises independent of the particular areas they have chosen to engage in.

The process of natural selection seems to have been particularly strong until 1975–76, when a considerable search for new areas of likely commercial success took place, with many new firms being created and many failing. Between 1969 and 1975 there was no significant net increase in the number of industrial establishments engaged in research and development (table 8.3)—there remained about 200 throughout the period—but there was considerable change in which firms were so engaged over the period in question. From an increase of 50% in the numbers of particular establishments engaging in research and development between 1969 and 1970—a rise from 210 to 308—we see an almost steady decline until 1975, although there was an increase between 1971 and 1972. Only from 1976 do we observe a continuous increase in the number of firms engaging in research and development (table 8.3). Within biomedical electronics, of the eight firms active in the field until 1973, two firms did not survive, and a third, while formally surviving, suspended practically all activities until a few years later; two other firms, while not disappearing, left the biomedical electronics instrument industry altogether. Between 1969 and 1975 —a period when the number of industrial establishments engaging in research and development remained constant—the share of the subsidy for research and development granted to electronics firms increased from 46% to 60% of the total subsidies for research and development granted to all firms. This change in the distribution of funds for research and development is a result of the process of natural selection.

There are two additional points worth mentioning in relation to the feasibility and efficiency of natural selection among firms and fields. Natural selection may be more or less efficient in generating firms of high quality and areas of high profitability. The existence of a pool of scientifically and technologically trained persons and the high quality of the higher educational system of the country enhanced the efficiency of the process. Similarly, the efficiency of the process was enhanced by the possibility of providing substantial financial support for research and development.

The natural selection of entrepreneurs or firms may be possible in some areas with divisible technologies, that is, in areas where small firms may acquire technical and economic efficiency, but it may not be possible in others with indivisible technologies. It is impossible within steel or petrochemicals, where a small country can maintain at most one plant. Certain areas within electronics, for example, have an advantage over "basic" industry with respect to their potential contribution to the economic growth of developing countries.

### Entrepreneurial learning

Since even the ultimately successful firms seem to have experienced considerable difficulties, including commercial failures, at the early stages of their existence, it is likely that the surviving firms are, on average, "fast learners" rather than that they have, by chance, selected at the outset a "good" program in research and development. It is possible that the initially excessive optimism of the eventually successful firms in some areas was a general phenomenon; it is also possible that their initially excessive optimism was a function of similar deficiencies in their understanding of the process of innovation and of the conditions for commercial success in innovation. A case study of the biomedical electronics industry suggests that the technological entrepreneurs at first thought that research and development were sufficient for success and that they could dispense with detailed, realistic knowledge of the market for their products. Thus overoptimism, at least in that segment of the electronics industry, arose from an underestimation of the importance of the market and of the marketing techniques necessary for commercial success (Teubal and Spiller 1977; this view is also consistent with the conclusions of project SAPPHO: see Science Policy Research Unit 1972).

This initial misconception of the determinants of successful innovation implies something about the nature of "entrepreneurial learning." In general terms, learning means both becoming aware of the need for taking account of the market and marketing requirements of new products, and becoming proficient in the assessment of the market and in marketing the new products. In biomedical electronics, this has involved, among other things, understanding the complex relationships between research and development and marketing. From a view which stressed the interchangeability of high quality in research and development with marketing knowledge and skill because the "product is so good it sells itself," entrepreneurial learning entailed acquiring a better understanding of the fact that the relationship between the two may be one either of interchangeability or of complementarity, depending on the character of the products arising from research. Thus, when higher quality or increased effort in research and development leads to products which are more novel from the point of view of the user, marketing requirements are almost inevitably greater. This is because of the greater difficulty in determining the nature of user needs and because the users must learn for themselves the functional efficiency of the product and "accept" it because of that. Only when high quality or more research and development leads to greater functional efficiency of products already standardized and well known to users—for example, via the enhancement of capabilities of data processing and display—might marketing efforts decline.

One implication of the importance of the marketing constraint on the commercial success of new entrepreneurs is that the conditions needed for realistic development of products include both technical feasibility and relatively low marketing requirements. This restricts the permitted degree of "product novelty"—from the users' viewpoint. Thus, the limited capital available to new entrepreneurs probably determined the commercial failure—at least within the field of biomedical electronic instruments—of new products about which there was much uncertainty about the particular character of users' needs and the way in which the products worked. (The latter is especially relevant for users prior to their decision to purchase the product.) These products are generally novel products to the users. In fact, between 7 and 13 of the 18 failures among the firms manufacturing biomedical electronic instruments occurred where the products were entirely new to their prospective users; the other important category of failure occurred in firms whose products (the exact number depends on the definition of "new") had no significant superiority in functional utility over competing products, even though they were offered at lower prices (Teubal, Arnon, and Trachtenberg 1976; Teubal and Spiller 1977). New enterprises in industries of high technology, even outside the production of biomedical electronic instruments, are more likely to be successful when they launch new products which are "known" to users and the main functions of which have been standardized. This conclusion applies to a wide variety of goods, especially industrial goods and especially those where a wrong choice on the part of the user may cause considerable damage—pacemakers, aircraft, and computers for process control are extreme cases of such goods.

From the point of view of the national economy rather than from that of the individual entrepreneur, however, a range of projects in research and development should include some "novel" and some risky ones. This is increasingly justified as the sector producing high technology grows and as groups of larger firms begin to emerge. The relative share of such projects in the early years of the system should not exceed, in my opinion, a very small percentage of the total.

The kind of entrepreneurial learning we are dealing with is that of new entrepreneurs with small firms in the "scanning stage"—a stage where the initial knowledge, skills, and capacities are used to search for a product on which they may base their subsequent growth and profitability. We are not dealing with subsequent stages in the growth of the firm, such as those associated with the establishment of manufacturing facilities, or with subsequent stages associated with reorganization, formal financial controls, and formal planning procedures. Learning of varieties other than that described above may be associated with these other stages. The types of areas associated with the type of entrepreneurial learning in question are those with dynamic innovation of products where a "product decision" which is not at all obvious precedes a decision on process.

Thus we are not concerned here at all with learning of process industries, which is usually analyzed in the literature on the transfer of technology, i.e., learning associated with the selection of technology and with the installation and "start-up" of productive facilities.

### Signs of Maturity: Emergence of a Group of Large Firms

As a consequence of natural selection, a group of larger firms emerged which, having succeeded in the past, were able to undertake more significant projects of research and development on possibly more complex varieties of existing products or in order to launch completely new classes of products. The risks and the opportunities associated with these firms in their postscanning, "growth" stage are probably of a completely different character than those confronting recently founded firms.

A basic feature of a firm entering its "growth" stage should be its past commercial success in innovation. This is proof of effective entrepreneurial learning and hence evidence of a capacity to innovate successfully. One possible measure is the size of the firm's budget for research and development. Taking $120,000 (at 1974 prices) as the "threshold level," the numbers of firms whose yearly budgets for civilian research and development exceeded this figure were 5 in 1974, 30 in 1975, and 40 in 1976.[4] Similarly, in 1975 there were eight electronics firms engaged in research and development and five chemical firms so engaged with sales over $10 million annually. This information, even though incomplete, testifies to the emergence of a fair number of firms with a capacity to undertake larger and more significant projects in research and development. My hypothesis, still unproven, is that the contribution of these firms to the growth of all high-technology industry was very significant. In electronics, for example, while the number of firms engaged in research and development increased from 24 to 53 between 1970 and 1976, their average budgets for research and development increased by 70% from $250,000 to approximately $470,000 (at 1971 prices). This includes one very large firm spending several million dollars on research and development. Other sectors had even higher rates of growth in average size, although electronics remained the sector with the highest average of expenditure research and development in 1970 (Pakes and Lach 1982).

The policy for the promotion of industrial technology was adapted to the emerging maturity of the sector by the launching in 1976 and 1977 of a new scheme for the support of research and development entitled the National Programme. This scheme was intended for "proven" firms only—i.e., those which had passed through the scanning stage—and which wished to undertake significant programs in research and development with high risks. The National Programme was not open to new firms or to existing but still unproven firms. Thus, in contrast to the scheme which supplied 50% of the funds used for research and development but did not discriminate among firms, this new scheme did discriminate while maintaining substantial "neutrality" with respect to branch of industry and class of product. A proven firm was required to be willing to undertake a very significant program of research and development such as the launching of complex new products which needed the incorporation of new technological inventions and possibly collaboration with scientists at universities, and which would capitalize on accumulated intangibles such as knowledge from research and development, experience in marketing, and the reputation of the firm. Thus, the scheme, while still neutral in some respects, was oriented towards growth through the reinforcement of existing success, moving on to new and more complex kinds of products. This promotional plan, which never provided more than 30% of the total of all grants made in a particular year, probably contributed to the growth of industry after 1976 in a significant way. Meanwhile, the National Programme has disappeared.

4. Office of the Chief Scientist, (1976). These are firms receiving support from the office and presumably mainly involved in civilian industrial research and development.

Most of the firms which benefited from it now have direct access to the Israeli and international capital markets. They have probably entered a phase of "maturity" where they control at least a small percentage of the world market of the products they develop and sell.

### Revealed Comparative Advantages: Toward the Understanding of the Dynamics of Growth in Industries of High Technology

Israel has experienced a dramatic shift in the structure of its comparative advantages during the 1970s; it is a shift which favors a group of industries which may be called "sophisticated" (table 8.5). The share of sophisticated industries in total exports, excluding diamonds, increased from 40% in 1970 to 66% in 1980. It is less dramatic if one examines the change in the share of this group of sophisticated industries in the total industrial output; still the average rate of growth in real output of this group was almost 50% higher than that of traditional industries: 6.1% against 4.2%.

I believe that an analysis of the effects of the technology promotion policies of the Office of the Chief Scientist should be related to a description and analysis of changes in the structure of exports and output, at least of the research-intensive, or sophisticated, industries. Otherwise we may fail to understand the process that led to increased exports. For example, it seems that the increased exports stimulated by the grants for research and development were composed of products not available in earlier years and, in many cases, of products which would not have been developed and produced at all in the absence of such a support scheme. The Office of the Chief Scientist apparently facilitated the transition to more complex and sophisticated products, and only through these were the increased exports achieved.

Our objective is thus to map the evolution of the comparative advantages of Israeli industries using high technology since 1967. One possibility is to focus on industrial branches or sectors and their changing share in the total exports or output of the group of sophisticated industries. This would give us a first, although crude, indication. The branches whose share in the total exports of the sophisticated industries increased between 1970 and 1980 were transportation equipment (in which the share increased from 6% in 1970 to 17% in 1980), electronics (where the shift was from 8% to 12%), and miscellaneous manufacturing. The share of all other sectors within this group of industries declined, especially rubber and plastics but including chemicals, metal products, and machinery. Electronic technology is intimately involved in all three of the industrial sectors where the share has increased ("optical equipment" is classified under "miscellaneous") (table 8.6).

These trends correspond with the trends in the share of total expenditures

Table 8.6. Exports of various branches of sophisticated industries: 1970–1980

| | 1970 | | 1975 | | 1980 | |
|---|---|---|---|---|---|---|
| | Dollars (current U.S. in millions) | Share (%) | Dollars (current U.S. in millions) | Share (%) | Dollars (current U.S. in millions) | Share (%) |
| Rubber and plastics | 24 | 15 | 45 | 8 | 128 | 6 |
| Chemicals | 53 | 33 | 183 | 34 | 648 | 30 |
| Metal products | 28 | 18 } 29 | 103 | 19 } 25 | 334 | 15 } 23 |
| Machinery | 18 | 11 } | 31 | 6 } | 171 | 8 } |
| Electronics | 13 | 8 | 98 | 18 | 263 | 12 |
| Transport equipment | 9 | 6 | 40 | 7 | 372 | 17 |
| Miscellaneous | 13 | 8 | 30 | 6 | 227 | 10 |
| Total | 158 | 100 | 530 | 100 | 2143 | 100 |

*Source:* Central Bureau of Statistics, *Statistical Abstract of Israel* (Jerusalem, various years).

for civilian industrial research and development, both that supported by the Office of the Chief Scientist and others, spent in the electronics and scientific instruments industries. This share increased from 45% in 1970 to 59% in 1976, which is an increase in share of almost 30% (Pakes and Lach 1982). Similar trends can be found in the share of grants by the Office of the Chief Scientist supporting projects in research and development in electronics: a moderate increase in share occurred in the period from 1967 to 1976 and a very significant increase between 1977 and 1981. The data, however, are incomplete and the two subperiods are not comparable. The substantial increase in the share of the grants made by the Office of the Chief Scientist to electronics after 1976 is a result of natural selection within the context of a policy deliberately favorable to this technology.[5]

This description of the changing importance of the various industrial branches in the total exports of sophisticated products is not sufficiently rich in detail to permit us to trace the effects of the various schemes of the Office of the Chief Scientist for the support of research and development. The main reasons are that support was aimed at particular innovations rather than at individual industrial branches, and that an important effect of the support was indirect, i.e., via the stimulus received to launch other related innovations in the future, over and above the particular innovations being supported.

In view of the deficiencies of using industrial branch or sector as the unit of analysis, I suggest considering products or innovations as the starting point

5. Most of this information is obtained from publications of the Ministry of Commerce and Industry, Office of the Chief Scientist.

and classifying them into a set of areas where each area is defined by one, two, or all of the following characteristics: class of product, field(s) of science and technology, and users. Successful innovations till 1975 (Ministry of Commerce and Industry 1976*a*) can be provisionally grouped by certain characteristics which have some relevance for the explanation of success. For example, an important factor for commercial success in the category "inputs to agriculture" has been the existence of innovative farmers who are "sophisticated users." More detailed knowledge of the classes of innovation within this group may eventually lead us to consider for example two "areas": chemical inputs to agriculture and all others. For the time being, however, and in order to provide as much concreteness as possible to our discussion, we will group the innovations—all of which benefited from the grants for research and development from the Office of the Chief Scientist—as follows: The pattern in agricultural innovation is one of supplying the local market first and then exporting. The existence of innovative farmers is significant here; some products were not new although the processes of producing and "using" them were novel. Success depended on interest and involvement of "users." The products comprised components and systems for irrigation, including those related to a novel drip-irrigation technique; herbicides, pesticides, and products to protect plants; fertilizers based on local natural resources; poultry; veterinary products; etc. Most exports in this group, such as fertilizers, are produced by the chemical industry; others are produced by the electronics industry, e.g., devices for irrigation control. Innovations in mechanical technology, such as orange-pickers, have multiplied in recent years and are already being exported.

Certain innovations arise from or occur in more than one field of science or technology. They generally involve the application of electronic technology to some other field. In medicine, the application of this technology has led to nuclear medical-diagnostic instruments, coronary care units, and laser-based operating instruments; in agriculture, it appears in computerized irrigation control; in textiles, dyeing machines operated and controlled by microprocessors and computer-aided design systems for knitting and printing. Most exports in this group come from the electronics industry. Starting in the mid-1970s, numerically controlled machine tools were developed and exported. Short lines of communication enable a small country like Israel to perform well in these areas. The influence of various fields of technology converging in a particular innovation or spreading from it often depends on the existence of sophisticated users, such as in medicine or agriculture.

The availability of large supplies of agricultural products opened up new opportunities for processing. Some of these involved innovations in products or processes, e.g., frozen and dehydrated citrus products and the extraction of sweet substances from orange peel. Presumably, exports from these innovations appear under the food and possibly chemical industries. Innovations in civilian industry occur in situations in which Israel's advantage derives from prior investments in defense, e.g., light aircraft and command control systems. Specific local needs sometimes engender innovations such as solar energy products and desalinization plants. The desire to find new ways to use local natural resources such as bromine and magnesium also stimulates technological innovation. The result of all this is that there has been a growing number of new products and processes of increased sophistication. Many of these products were not available before 1974 (Ministry of Commerce and Industry 1974, 1976b, 1977). Between 1971 and 1977 many new products were added. In communication equipment, there has been the addition of telephone exchanges, microwave radar, airport tower communication equipment, remote control and data transmission systems, etc.; in medical instruments, gamma cameras, CAT scanners, coronary care and intensive care units, and surgical laser devices; in electro-optical systems and other instruments, of CAD systems first for the textile and then for the printing industry, "fast-Fourier processor," airborne instruments, etc.; numerically controlled sewing machines, electronic packaging, numerically controlled machine tools; in computer equipment, a variety of terminals and alphanumeric displays.

These innovations illustrate some of the consequences of the activities of the Office of the Chief Scientist in the support of research and development. Yet although the innovations benefited to some extent from governmental support, and in some cases it is clear that such support had a very strong impact, the precise mechanisms at work are still unknown. This may require assembling chains of interrelated innovations over a period of time.

## Concluding Remarks

In contrast to other systems for the promotion of scientific industrial technology, the Israeli system has worked directly through individual industrial firms. The location of the Office of the Chief Scientist in the Ministry of Commerce and Industry has permitted the restricted application of the allocative criterion of "commercial prospects" rather than that of scientific originality; the latter is not disregarded but is considered only if the former criterion is satisfied. The system operates through grants to specific research and development projects rather than attempting to stimulate innovation through selective procurement of industrial products benefiting local firms. This procedure, therefore, differs from that of other countries such as France, and it is probably best adapted to encourage the emergence of new, usually young, technologically sophisticated entrepreneurs. It remains to be seen whether the use of grants for research and development to promote industries using high technology is most effective after a stage where such promotion has already been furthered by procurement policies, or even whether it is necessary where there are well-developed capital markets.

The fruitfulness of the Israeli system has derived in part from the formal neutrality of it support to research and development at least during its first decade of existence. The practice has not been unqualifiedly neutral with respect to innovation, and of course it has favored those firms which were already interested in using the results of their own research and development in order to make innovations in products and processes. It has been a neutrality intended to promote sophisticated technological innovation through research and development; it has not been neutral with regard to traditional as opposed to scientifically sophisticated technology, or with regard to the latter as opposed to imported technology. It has been very partisan indeed in the promotion of local scientifically sophisticated technology, but it has practiced neutrality as a means to that end, at least during the ten years following 1967. This policy of neutrality has been successful in enabling new firms to find their way and to prosper through research and development. Can it be adapted to new circumstances to continue to further technological innovation?

### References

Arnon, N. 1978. "Principal Developments in Science and Technology during the First 30 Years of the Nation." (In Hebrew.) *Maarajot* (April).

Dudai, Y. 1974. *Scientific Research in Israel.* (In Hebrew.) Jerusalem: National Council for Research and Development.

Hershkovitz, S. 1980. *Government Allocations to R & D in Israel, 1976/77–1978/79.* (In Hebrew.) Jerusalem: National Council for Research and Development.

Irvine, J., B. Martin, M. Schwarz, K. Pavitt, and R. Rothwell. 1982. *The Assessment of Government Support for Industrial Research: Lessons from a Study of Norway.* Lewes: Science Policy Research Unit, University of Sussex.

Katz, Shaul, and Joseph Ben-David. 1975. "Scientific Research and Agricultural Innovation in Israel." *Minerva* 13:152–82.

Magaziner, I., and T. Hout. 1980. *Japanese Industrial Policy.* London: Policy Studies Institute.

Mansfield, E., J. Rapoport, J. Schnee, S. Wagner, and M. Hamburger. 1971. *Research and Innovation in the Modern Corporation.* New York: Norton.

Ministry of Commerce and Industry. Office of the Chief Scientist. 1974. *Directory of Science-based Industry.* Jerusalem.

Ministry of commerce and Industry. Office of the Chief Scientist. 1976a. *Industrial R & D.* (In Hebrew.) Jerusalem.

Ministry of Commerce and Industry. Office of the Chief Scientist. 1976b. *Directory of Science-based Industry.* Jerusalem.

Ministry of Commerce and Industry. Office of the Chief Scientist. 1977. *Directory of Science-based Industry.* Jerusalem.

Pakes, A., and S. Lach. 1982. *Civilian Research and Development Activity in Israeli*

*Industry: A Look at the Data.* Jerusalem: Maurice Falk Institute for Economic Research in Israel.

Science Policy Research Unit, University of Sussex. 1972. *Success and Failure in Industrial Innovation: Report on Project SAPPHO.* London: Centre for the Study of Industrial Innovation.

Teubal, Morris, Naftali Arnon, and Manuel Trachtenberg. 1976. "Performance in Innovation in the Israeli Electronics Industry: A Case Study of Biomedical Electronics Instrumentation." *Research Policy* 5:534–79. (Chap. 1 in this volume.)

Teubal, M. and Pablo T. Spiller. 1977. "Analysis of R & D Failure." *Research Policy* 6:254–75. (Chap. 2 in this volume.)

Westphal, L. 1982. "Fostering Technology Mastery by Means of Selective Infant-Industry Protection." In *Trade, Stability, Technology, and Equity in Latin America,* edited by M. Syrquin and S. Teitel, 255–79. New York: Academic Press.

Some of the issues of this paper have been raised in a previous work entitled "The Science and Technology System of Israel: An Overview with Special Emphasis on Industrial Research," October 1982. I appreciate the comments and help received from J. Katz, S. Katz, H. Pack, A. Pakes, K. Pavitt, C. Perez, E. Shils, L. Soete, B. Toren, and from participants of seminars given at the Science Policy Research Unit, University of Sussex, and at Bureau d'economie theoretique et appliquée, University of Strasbourg.

# 9 Innovation Policy in an Open Economy
## A Normative Framework for Strategic and Tactical Issues

## Moshe Justman and Morris Teubal

### Introduction

The approach conventionally used by economists to justify government support of industrial innovation is the so-called market failure approach. According to this approach, a disparity between the profitability of an activity for society as a whole ("social" profitability) and its profitability for private firms may lead the economy to allocate its resources in a way which is suboptimal. This approach was first developed in the early work of Nelson (1959) and Arrow (1961) but was later sharply criticized by those who followed in the wake of Schumpeter's evolutionary view of the innovation process (including Nelson himself in his more recent work). Our objective in this paper is to approach a synthesis of the market failure justification of government support of industrial innovation with its Schumpeterian critique within the concrete context of policy analysis.

Arrow (1961) identified three basic characteristics of information and of its production that cause market forces to underinvest in activities leading to innovation: externalities, indivisibilities, and uncertainty. His arguments are largely derived from a simplified view of the innovation process, one that focuses on the firm as a producer of information, neglecting for the most part the structural aspects of R & D promotion and the international dimension of competition in high-technology industries. Nonetheless, his work has spawned a large body of theoretical analysis.

However, with the notable exception of Nelson, Peck, and Kalacheck (1967), there has been little application of these principles to concrete policy issues. Nelson et al. provide an illuminating description and analysis of United States government support of R & D in the early 1960s which duly stresses the role of market failure in explaining the pattern of support across broad areas, industries, and technologies, while recognizing the importance of other factors as well. It should be noted, however, that the special characteristics of the

United States economy at the time—the world economic and technological leader possessing the largest selection of big and mature firms—limit the direct relevance of their analysis to other countries. Indeed, the circumstances of the United States economy have since changed markedly. Open economy issues, which played a very minor role in the context of Nelson et al.'s analysis, are today paramount in high-technology industries. And there is increasing recognition of the role of industrial R & D support systems in developing a national capability of industrial innovation. These aspects introduce additional dimensions to the market failure analysis of government policy towards innovation (see also Justman 1984). Concomitantly, Nelson et al. do not provide an analysis of broad schemes of direct government financial assistance to industrial R & D of the type gradually being implemented in Europe today (Rothwell and Zegveld 1981), and which has been implemented in Israel since 1967 (Teubal 1984).

The evolutionary perspective of economic change (see Nelson and Winter (1982) and other recent work by Nelson (1983, 1984)) have emphasized more than previously the limitations of the market failure approach, or at least of the simple form of static market failure analysis that has been common in the past (based on fully rational, profit-maximizing firms functioning in an environment with no fundamental uncertainty and substantial equilibrium). Clearly, the basic assumptions and perspectives of the market failure approach are still not good enough, and some major policy issues—such as how to assure a plentiful supply of Schumpeterian entrepreneurs—cannot yet even be addressed by the conventional framework analysis. Moreover, the actual pattern of government support is shaped by other forces beyond "market failure" (Nelson 1983). There are inherent constraints on governments: rivalrous competition and incomplete access to critical information required for cooperative research efforts involving specific designs and whether or not a scientific base to innovation exists are examples.

While recognizing these and other limitations of conventional analysis, we contend nonetheless that the systematic attempt to identify activities associated with gaps between private and social profitability represents the analytical basis for innovation policy, *and in this sense*, further development and adaptation of the market failure approach are warranted. This development, however, need not be based on equilibrium analysis; on the contrary, its perspective may be explicitly dynamic and based on attempts at a detailed observation of real-world phenomena. Thus our paper is not meant to supplant the large body of economic analysis in the tradition of Schumpeter, which has come to have a profound, if not dominating, influence on both the theory and practice of innovation policy. Rather it is meant to complement it. Entrepreneurs are without doubt an essential factor in achieving technological progress, and rationality in a dynamic, changing environment is without doubt bounded. But the right economic incentives can help entrepreneurs produce desired results, and rationality, though bounded, is most certainly not absent.

The fact that market failures are ubiquitous in relation to innovation (Nelson and Winter 1982) only reinforces our argument. Since it is no longer enough to identify a market failure in order to conclude that a particular action is needed, there is substance to the statement that the existing mode of reasoning cannot carry policy analysis very far. At least an explicit ranking of market failures in terms of the benefits and costs of remedial action is required. Theory should then attempt to help us identify the various types of market failure which are likely to appear at the various stages in the development of a high-technology industrial sector; determine their likely relative importance; and recommend the kinds of policies to be applied. The recommendations should in turn explicitly take into account the constraints on government action. In short, while we agree that a normative analysis of innovation policy may be hindered by simple, static market failure analysis, there is room for other types of such analysis that recognize the existence of phenomena such as "bounded rationality and slow-moving selection."

This general approach is applied in the present paper in the analysis of government policy towards industrial innovation, which we take to mean both policy towards industrial R & D and policy related to the application of R & D results. Consequently, we include here only those aspects of science and technology policy which affect the infrastructure supporting industrial innovation. And we do not explicitly consider the diffusion of innovations (although the diffusion of technologies is part of our infrastructure development) or the policies for the development of technological capabilities which are not *directly* related to R & D capabilities or to the relevant infrastructure.[1] It is in this specific context that we undertake further development of the market failure approach to innovation policy. Besides the aspects of market failure usually presented in the literature, we will relate to the following ones: open economy considerations (issues such as the incidence of externalities, the special "balance" problem between infrastructure support and support to specific innovations, etc.), externalities derived from experience and related to the development of innovation capabilities (rather than Arrow's "template" externalities), and an explicit distinction between industrial R & D and the *application* of R & D results to the production and marketing of new products and processes.

Throughout the analysis we will make the critical distinction between support of infrastructure and support of specific innovations (or R & D projects).

---

1. For a case study of the development of a wide spectrum of technological capabilities, including but not emphasizing R & D capabilities, see Westphal, Kim, and Dahlman 1984.

This distinction is based on indivisibilities and may have far-reaching implications concerning the optimal combination of public and private decision making in resource allocation. It is reflected in what we believe to be a most useful distinction between tactical and strategic instruments. Whereas the former are limited in scope and hence lend themselves to the partial equilibrium analysis of bureaucratic management, the latter will often have far-reaching effects that must be considered in the more general context of national priorities, which must ultimately be decided at the political level.

Finally we would like to emphasize that the present analysis is concerned primarily with necessary conditions for government intervention in various circumstances. They are also sufficient conditions only if government can do better than the free market in the pertinent dimensions. Clearly this need not always be the case. The absence of a theory of government failure as developed as the economic theory of market failure does not imply the absence of such failure. The ability of government to execute a desired policy as conceived is without doubt an important consideration in the adoption of that policy. Nonetheless, difficulties of implementation of the various innovation policies considered in this paper are generally beyond its scope.

In the following section we set out, in brief, the basic economic principles from which R & D policies must draw their theoretical support. We then go on to consider, in subsequent sections, the two main components of such policy: "strategic" support for infrastructure development, and "tactical" support for specific R & D projects. The paper concludes with a brief summary of the policy implications of the analysis.

## The Economic Principles

The point of departure for our analysis is that market failure is a necessary condition for public intervention in the innovation process: there must be some discrepancy between private motives and public goals to warrant intervention. Such a discrepancy may stem from a variety of sources, which we find convenient to group here under four headings: externalities stemming from the low social cost of transferring information; uncertainty; economies of scale surrounding the innovation process; and strategic considerations stemming from imperfections in world markets.

Each of these may prevent socially worthwhile economic activities in the R & D process from offering an expected return that would justify private investment; or it may be that levels of R & D investment by private enterprise may be lower (or higher) than is socially optimal. In such cases public intervention is warranted if it is able to modify the behavior of private enterprise and thereby increase the net benefits stemming from innovative activity to society as a whole.

### The low social cost of transferring information

Analysis of market failure in the allocation of resources for R & D has followed largely in the wake of Arrow's (1961) seminal work in stressing the low social cost of transferring information as a primary source of such failure, and one that is inherent in the innovative process. Moreover, it leads to a conflict of effects that cannot be resolved completely.

The low social cost of transferring information implies that once information is produced, it should be distributed freely. This applies both to the final product of the innovative process and to intermediate results, so as to prevent both insufficient dissemination of inventions and wasteful duplication of effort. From another perspective, this same phenomenon also implies that the gains from innovation cannot be appropriated in their entirety by the innovator, and this may lead to underinvestment in research and development.

Clearly, dealing with these conflicting effects simultaneously raises serious difficulties. If incentives are strengthened by increasing appropriability, e.g., through the patent system, then the diffusion of innovations seems bound to suffer, and wasteful duplication is also likely. Conversely, increasing the dissemination of the results of innovative activity seems to imply less payback for the innovator and consequently fewer incentives to innovate.

Previous analysis has largely dealt with characterizing the trade-off implicit in this quandary within the context of a closed economy. While recognizing the importance of this approach, we go on to consider its extension to an open economy. In an open economy the incidence of externalities must be taken into account, and this may well tip the balance in one direction or another. This issue is taken up in greater detail in a later section, where R & D subsidies are considered (see also Justman 1984).

### Uncertainty

Arrow (1961) also provides a basis for analyzing the implications of uncertainty for research policy. Indeed, on a general level there seems little to add to his insights on this issue even today.

While emphasizing the importance of capital markets for shifting risks, Arrow also stresses the inherent limitations of such markets. The same instrument that shifts risk from innovator to investor may also dull the innovator's profit motive. Imperfect monitoring of the investor by the innovator and the presumed divergence of purpose between the two must then lead to underinvestment. Arrow referred to this as the "moral factor," and more recent usage terms it the "moral hazard."

Moreover, capital markets are generally imperfect. In some cases these imperfections may be rectifiable through government intervention; e.g., when prohibitive fixed costs prevent the creation of a market, some government subsidy may be desirable to allow the market to function.

### Economies of scale

While the moment of invention is often characterized by few economies of scale, it is often surrounded, both in the preparations that lead up to it and in its eventual development and dissemination, by scale economies of a large order.

When such scale economies can be absorbed by the individual firm, through vertical integration, no issue of market failure arises. But when they transcend the needs of the individual firm, it may well be the case that the social benefits associated with some of these various supporting elements exceed their cost, even though they may not yet be profitable for the private entrepreneur. These issues are dealt with at greater length in a later section, where the infrastructure of innovation is considered.

### Strategic intervention

The opportunity for strategic intervention on the part of national government arises in international markets in which competition is imperfect and supranormal profits can be earned. Division of these profits between market competitors in such instances cannot be determined a priori on economic grounds but depends on the strategic interaction between them. Brander and Spencer address these issues in a series of papers (e.g., 1985) in which they make the point that by subsidizing R & D, governments can ultimately reduce the marginal cost of production by domestic manufacturers and consequently increase their market share and profits. Other forms of output subsidy, direct or indirect, might serve the same purpose. However, R & D subsidies have the advantage of being applied at the earliest stage of the production process. They are thus less likely to elicit retaliatory responses than are more direct forms of output subsidy. Further analysis of strategic intervention can be found in a later section.

### Infrastructure

Technological progress does not begin and end with the innovating firm. Rather, both the initial innovation and its subsequent development and diffusion are strongly affected by external conditions beyond the scope of the individual firm. It is these conditions which provide the infrastructure for successful technological innovation.

In this broad sense the infrastructure of innovation comprises a wide range of elements embodied in a variety of forms and supporting different stages of the extended process of innovation and diffusion. These would include both traditional elements of physical infrastructure such as power, transportation, and communications, as well as less tangible elements embodied in human capital, such as technical and scientific skills. They have in common economies of scale in their production and supply that transcend the needs of the individual firm so that they cannot be vertically integrated with the innovating firm and must be supplied externally.

Table 9.1 presents the essential components of the infrastructure, organized in two dimensions: according to the medium in which they are embodied (human capital, organizational structure, physical capital) and according to the particular function in the innovation process that they serve (finance, research and development, manufacturing,[2] marketing). The selection is not comprehensive nor is its classification definitive, but it provides a good idea of the types of phenomena with which we are concerned in this section. It also provides a framework for an integrated analysis of the infrastructure in its entirety. Such an analysis can help rectify imbalances in the infrastructure by diverting development efforts to its weakest links.

In the following subsection we consider the theoretical economic justification for public intervention in support of the infrastructure and the conditions under which such intervention is justified. We then go on to apply this theoretical perspective to each of the economic functions which different elements of the infrastructure must serve. The section concludes with some comments on the problem of achieving balanced development of the infrastructure.

### Some theoretical considerations

Spence's (1976) analysis of product selection under monopolistic competition develops the theoretical basis for government support of infrastructure development on neoclassical grounds. This stems from the discrete, i.e., "lumpy," nature of investment in infrastructure and the economies of scale underlying this lumpiness. These preclude its being supplied competitively and raise the possibility of a substantial disparity between the private and public criteria for providing the goods and services that make up the infrastructure.

The principal condition for such a disparity to exist with regard to a particular investment is that the scale of the investment transcends the scope of the individual firm, so that internalization of all the benefits of investment through vertical integration is not possible. In addition, the demand curve for infrastructure services (or goods) must be downward sloping, reflecting differ-

---

2. Strictly speaking, manufacturing is not part of the innovation process, but the exploitation of local innovations by domestic industry is clearly a principal goal of all innovation policies insofar as they are incorporated in an overall industrial policy. See Williams 1983.

Table 9.1. Components of the infrastructure

*1. Finance*
  1.1  R & D stage
       H - Financial assessment and management of R & D projects
       O - Venture capital market, government funding mechanisms
  1.2  Implementation
       H - Financial management of implementation stage (high risk, growth)
       O - OTC market, international correspondence network, government funding
           mechanisms

*2. Research and Development*
  2.1  Access to state of art
       H - Scientific and technical knowledge
       O - Universities, research institutes, links with foreign research institutes
       P - libraries and data bases, basic and generic research facilities (e.g., laboratories)
  2.2  Experimentation
       H - Inventive abilities, Science and Technology
       O - Applied research groups, testing facilities
       P - Experimentation facilities
  2.3  Design
       H - "Design or innovation capabilities"—all of the above plus market/economic
           orientation experience
       O - Innovating firms, market research organizations, subcontracting of specialized
           design work
       P - Design facilities

*3. Manufacturing*
  3.1  In the R & D process
       H - Prototype manufacturing skills
       O - Firms geared to specialized high-skill products of type needeed
       P - Machinery for producing prototypes
  3.2  Exploiting the fruits of R & D
       H - Manufacturing skills in relevant industry, capacity for quick study
       O - Firms geared to producing/using new products and custom-made components
           (flexible, open); university and skilled manpower training facilities
       P - Manufacturing facilities that can produce/use new products

*4. Marketing*
       H - High technology, marketing skills, reputation, national goodwill
       O - Trading companies, export boards, market research organizations; marketing
           training institutions
       P - Distribution and warehousing facilities.

*Note*: H = embodied in human capital; O = embodied in organizational structures;
P = embodied in physical capital.

ences in the value placed on these services by different buyers (or by the same buyer for different uses). Furthermore it clearly must not be possible to practice perfect price discrimination in providing infrastructure services, or to import perfect substitutes at no additional cost. Given these conditions, Spence's formal demonstration that the market selects against new products characterized by large fixed costs and low elasticities of demand can be applied to those products or services that make up the infrastructure of innovation.

These general arguments are presented in a static context in figure 9.1, which depicts the conditions of supply and demand of a new product in a closed economy under increasing returns to scale. For simplicity, increasing returns to scale takes the form of a fixed introduction cost $F$ and a constant variable cost $v$. Given the demand curve for the product, $DD$, the optimal supply for the good by the private investor is $X_0$ (we assume he is a monopolist who cannot discriminate among users), and the price charged is $P_0$. Private and social profits, $\pi_p$ and $\pi_s$, are given by

$$\pi_p = -F + abcd \qquad \text{and} \qquad \pi_s = (-F + abcd) + bec,$$

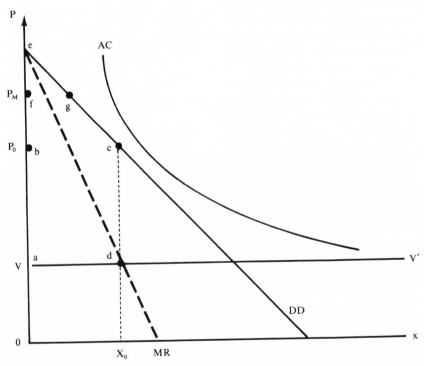

Fig. 9.1. Market failure under increasing returns: the product margin.

where $abcd = X_0 (P_0 - v)$ are private operating profits, and $bec$ is the additional surplus generated for users of the product (an external economy from introduction of the new product). When the condition

$$abcd + bec > F > abcd$$

holds, it implies

$$\Pi_s > 0 > \Pi_p.$$

Thus, supply of the product is socially but not privately profitable, suggesting a possible need for government intervention.

Clearly, if price discrimination can be practiced, profits will more closely approximate social gain; if perfect price discrimination is possible, there is no disparity between the two. And if the product or service can be imported, say at a price of $P_m$ on the diagram, then social gain will more closely approximate profits, the disparity disappearing when the import price is equal to or lower than the monopoly price.

In a dynamic context the demand curve will be moving outward in response to technological or economic developments. Supply of infrastructure services by the private sector then becomes profitable when the demand curve meets the average cost curve $(AC)$. Thus many investments that warrant government support at the early stages of industry development would later be undertaken through private initiative. But for many investments of this type the private initiative will come later than is socially optimal.

These general principles are applied in the remainder of this section, in analyzing the scope for government intervention in each of the components of the infrastructure.

### Finance

Innovation requires financial support on three possible levels. In the initial stages of innovation venture capital must be raised to finance the first efforts of learning and experimentation that characterize this phase. Then growth capital must be found to fund the implementation and initial diffusion of the innovation. Finally, if the potential market for the innovation transcends national borders, successful diffusion will require the support of international banking facilities.

Implications of these needs for infrastructure development are first and foremost organizational. A venture capital market and/or some government funding mechanism are required if capital is to be allocated for the early stages of innovation. An over-the-counter (OTC) market or parallel government mechanism is needed to finance the subsequent high-growth stage of initial diffusion. Foreign credit arrangements are required if international diffusion is to succeed.

Each such organization element involves considerable economies of scale in several respects: information needs, risk spreading, and administration. Moreover, foreign credit markets and credit facilities are rarely a feasible alternative for domestic innovators. Also, the high element of uncertainty inherent in the innovative process coupled with the moral hazard that arises from the asymmetry of information between innovator and investor would seem to preclude any comprehensive attempt at price discrimination. These conditions therefore appear to support the inclusion of financial markets among the elements of infrastructure for innovation for which government assistance is possibly justified.

Indeed, government intervention in the financial aspects of innovation has been observed in many countries in recent years. Such a function is served in the Federal Republic of Germany by the Ministry for Research and Technology (BMFT), which funds industrial projects aimed at enhancing the technological competence of German industry, generally on a mutual basis with the private sector (Nelson 1984). Spain's Center for the Development of Industrial Technology, a part of the Ministry of Industry and Energy, extends loans for R & D efforts within industrial enterprises (Rao and Weiss 1982). Israel's Industrial R & D Fund, administered by the Office of the Chief Scientist of the Ministry of Industry and Commerce, extends grants on a matching basis to industrial firms for the execution of R & D projects (Teubal 1984). These government programs augment private sector venture capital by financing the early stages of innovation through direct intervention. They are generally reinforced by preferential tax treatment for R & D expenditures aimed at stimulating private investment.

Government-sponsored institutions aimed at improving foreign credit arrangements are less common (though subsidized government credit for large foreign purchases is a common form of indirect export subsidy). They are more likely to be needed in the newly industrializing countries where the private banking sector has not kept pace with industry. Korea's Export Import Bank, which provides long-term credit for foreign purchase of Korean plants and turnkey projects, is a good example of such an institution which has developed to meet the special needs of local industry.

Beyond the institutional and organizational arrangements needed to meet the financial needs of technological innovation, indeed stemming from them, there arises a need for special skills or human capital. People must evaluate the needs and prospects of new R & D projects, undertake the financial management of the critical high-growth stage, and make international credit arrangements. Because of the moral hazard inherent in any external investment in human capital (Urban 1983), the supply of skilled manpower to meet these needs is again an element of infrastructure that possibly merits government support.

These training requirements can be met in several ways, applied in combination: through formal education, apprenticeship in the private sector, and apprenticeship in the public sector. The more technical aspects of financing R & D are better served by formal education with some limited stewardship; the more creative aspects require a larger share of learning by doing. The public sector has an important role to play in providing the relevant business education. And it can also provide opportunities for gaining practical experience in financing technological innovation, if it is involved in these activities. Indeed, accumulation of such experience may be an important consideration favoring direct government financing of R & D in the earlier stages of technological development.[3]

### Research and development

Research and development is, of course, at the heart of the innovation process. We have chosen here to divide it into two stages, each of which requires its special type of infrastructure support.

Generally, the first prerequisite for successful R & D is familiarity with the state of the art. While some innovations are so revolutionary so as to obviate all previous experience, this is, of course, very much the exception. Most innovations build on available scientific and technological knowledge. This is most clearly apparent in the electronics industry, where many innovations are achieved through a new assembly of existing components. But it is equally true in other fields characterized by high levels of innovation, such as pharmaceuticals, where innovative research without a sound scientific basis in chemistry is virtually unthinkable.

Familiarity with the state of the art is embodied first and foremost in human capital and as such is clearly an element of infrastructure because of the general imperfection of markets in human capital; investment in such markets is limited by an inherent asymmetry of information, which gives rise to an element of moral hazard. Moreover, such investments are likely to generate positive external effects.

On an organizational level ongoing support for this element of infrastructure is provided by research-oriented university departments and other research institutions engaged in basic and generic research as well as by the research departments of the larger innovating firms. Some measure of such research activity is essential for keeping abreast of developments in the state of

---

3. To some extent, government support of the financial infrastructure supporting the industrial R & D process may be viewed as support of an infant industry supplying an essential service (financial capital) to that process. That is, a learning process is involved in the establishment of an infrastructure element such as this, over and above the production indivisibilities reflected in our application of Spence's model of product selection.

the art, even if it yields few such developments itself. It is embodied also in the physical facilities (libraries, laboratory equipment, computers) that support these activities.

Public support for basic science is universally accepted as necessary and beneficial, though this is not so much the case for generic research. In this respect the Japanese experience is exceptional in its breadth as well as its measure of success. Japan's Ministry of International Trade and Industry (MITI) has played an active role in funding and orchestrating large-scale cooperative research efforts in various fields, including color television, very large scale integration (VLSI), computer design, and biotechnology. Japanese efforts were geared not to the development of specific products but to help Japanese manufacturers develop the capabilities to move in a variety of possible directions along a broad front (Nelson 1984; Peck 1983).

Generic scientific and technological knowledge includes broad design concepts, general working characteristics of processes, knowledge of materials characteristics, aspects of quality-control processes and testing methodology, and standards. Such knowledge is often not patentable; much of it is shared openly by scientists and engineers working in the field, whether they are located in universities, government, or corporate laboratories. Research systems involved in this kind of work fit in a niche between fundamental research and applied R & D for industrial firms (Nelson 1983).

The agricultural and medical sciences appear to have defined their niches appropriately, but public sponsoring of industrial generic research in the United States and in Europe has traditionally been lacking.[4] Recently, things have been changing, however, both in Europe and increasingly in the United States. For example, the newly established Microelectronics and Information Sciences Center at the University of Minnesota is conducting a joint research effort in the generic areas of integrated circuit design, microelectronics, and systems architecture. This effort involves both industry and government, with researchers having access to the facilities of the larger participatory firms, as is done in Japan (Tassey 1982).

While knowledge of the state of the art is a prerequisite for R & D, the essence of innovation is experimentation and design. These require specific scientific and technical skills, of course, but also some measure of creativity and a propensity for risk; and they require a commercial orientation that takes into account not only technological feasibility but also production costs and market demand. In short, the crucial stage of R & D requires the attributes generally associated with what has come to be known as the "Schumpeterian entrepreneur."

---

4. The pre–World War II experiments of the National Advisory Committee for Aeronautics with aircraft technology, which led to the DC-3, form one notable exception.

Much has been written about this essential element of the innovation process, with national success and failure attributed to its presence or absence, but little is known about its sources. Some importance must be attached to the basic psychological parameters of the economy, i.e., the "national mentality," which is affected, in turn, by the country's educational system, its class structure, and its social traditions. But it is far from clear how these influences act, and we have nothing to add that might help clarify their effects. They are beyond the scope of *economic* analysis.

However, we would argue that the supply of Schumpeterian entrepreneurs can be affected by economic policy. Entrepreneurs are, by definition, sensitive to profit opportunities, so that creating such new opportunities in technological innovation attracts potential entrepreneurs to try their hand. The number of people in the population with strong personal motivation, a capacity for finding creative solutions to difficult problems, and a propensity for taking calculated risks may very well not be susceptible to manipulation by economic means. But where these talents are applied—in the armed forces, in civil administration, in trade, in science or the arts, or in technological innovation—is influenced by economic incentives. One of the main objectives of Israel's innovation policy in its early stages was to exert just such an influence. Matching grants for research projects administered by the Office of the Chief Scientist of the Ministry of Industry and Trade had the beneficial effect of creating a pool of Schumpeterian entrepreneurs, who continue to play a key part in furthering the country's technological development. This effect clearly transcends the success or failure of the initial projects for which the funds were allotted (Teubal 1984).

Related to the generation of Schumpeterian entrepreneurs is the development through "learning by doing" of what may be termed experienced-based industrial innovation capabilities. "Doing" industrial innovation enhances these capabilities: both in regard to R & D itself (e.g., the desired combination of analogue and digital components in optical instruments) and to its relationship with subsequent steps of the innovation process (i.e., production and marketing). Moreover, it enhances the capabilities of new enterpreneurs to search for appropriate R & D projects. Thus, R & D project support, especially in the early stages of development of the high-technology industrial section, *indirectly* supports an important *infrastructure* element for successful innovation—the development of innovation capabilities. Private investors are unlikely to capitalize on all the social benefits derived from enhancement of these capabilities: the movement of individuals to other firms will generally carry not only specific information (Arrow's template externalities) but person-embodied general capabilities as well. Government subsidization should be regarded as an attempt to take into account the existence of both types of externalities. At the early stages of growth in a high-technology industry the stock of innovation capability in the economy is not yet sufficient. Therefore, work on almost any innovation will lead to a significant increase in these capabilities. Moreover, the general lack of experience will cause a large number of R & D projects to fail at this stage. Thus (relative to subsequent stages) a large share of the social benefits from innovation will take the form of externalities, and a large share of these involve enhanced general capabilities. At the "growth" or "maturity" stages, an increasingly important share of the external benefits from innovation will be of the run-of-the-mill kind, stemming from innovation-specific information flows and from additions to user surplus, with little effect on the pool of general innovation capabilities.

### Manufacturing

Manufacturing capabilities form part of the infrastructure of successful innovation in two different ways. In the R & D stage specific manufacturing capabilities are often necessary for experimentation and design. Models and prototypes must be custom-built in small quantities and meet stringent tolerance levels, often requiring unconventional materials, methods, or processes. These manufacturing services require close interaction with the innovator, which largely precludes their provision by foreign suppliers. Larger firms often solve the problem by developing in-house manufacturing capabilities, but this is clearly not a feasible option for smaller innovators. Moreover, the need for secrecy limits the ability of smaller firms to hire these services from their larger competitors.

The second stage at which manufacturing capabilities are crucial for success is in the implementation and diffusion of the innovation. Innovations contribute most fully to industrial development when they are embodied in products rather than sold or leased for foreign manufacture. This implies a need for a domestic manufacturing sector that can produce such products at competitive cost. Of course it also implies that R & D projects selected for government support must be chosen to accord with existing capabilities, as we elaborate in a later section.

Thus, for example, in electronics and other industries, a crucial role is played in this regard by an infrastructure of specialist producers of parts or components. The manufacturing of new products may be significantly hindered at the early stages of high-technology industrial growth by the absence of such a network of producers. In such situations, local firms will need to produce a much larger share of the product than best practice would indicate, causing higher costs and delays in complementary innovations, and placing them at a competitive disadvantage vis-à-vis foreign firms with better access to component sources. This was the case in Israel's electronics industry in the late sixties and early seventies, when local suppliers of custom-made printed circuits and other custom-made electronic components were absent. The indivisibilities in creating a capability to supply these specialized products or inputs are, prima facie, a reason for government support.

### Marketing

Marketing is a crucial element in diffusion. Economies of scale in marketing that transcend the needs of the individual firm and allow us to speak of a marketing infrastructure are most prevalent when diffusion of the innovation must transcend national boundaries and penetrate foreign markets.

In such cases a concerted effort by groups of firms, or a national effort, can capitalize on economies of scale in information on economic conditions and the legal and regulatory environment in foreign markets. Provision of general public information of this type is an accepted function of government in many countries. Beyond that, market research organizations as well as legal, commercial, and even language specialists are important private sector elements of the infrastructure of marketing information.

Agencies in foreign countries, for both sales and service, are another aspect of marketing that offers an opportunity for exploiting economies of scale. Japanese trading companies are, of course, the prime example of a private sector response to this opportunity, and one that has been replicated in many other countries. They offer the advantages of goodwill, access to potential buyers, distribution facilities, and technological capabilities that are generally beyond the scope of the single innovating firm, at least in the early stages of its growth.

Furthermore, penetration of an export market often depends on some element of national reputation. The label "Made in Japan" is an excellent example. Once a synonym for simple, low-cost products, it has since become a symbol of efficiency in production, quality control, and technological progress and as such is a great asset for Japanese manufacturers. A nation's commercial image is, of course, most strongly influenced by the quality and price of the goods and services it has to offer. It can be enhanced or formed more quickly through advertising campaigns, trade fairs and other forms of sales promotion, and penetration pricing. The first two of these measures are commonly the province of export boards (such as Israel's Export Institute). The last of these has apparently been subsidized by Japan's Ministry of International Trade and Industry (MITI) but is obviously more exceptional.

Finally, the public sector has a role to play in developing the human capital for international marketing. This it can do by promoting language and regional studies pertaining to export target areas as well as the relevant legal and commercial specialties. Beyond that, national economic representations in foreign countries can also add to the aggregate local experience of commercial conditions in these markets.

Balance

While each of the elements of the infrastructure described above must be examined separately, the success of technological development is determined by the overall performance of the system. Therefore, considerations of economic efficiency require that the balance between the different elements also be taken into account.

Ideally the educational system should produce just enough trained manpower, and no more than can be absorbed; generic research laboratories should spawn just the right number of new ideas and state-of-the-art techniques; manufacturing capabilities should be available to capitalize on new innovations; and the marketing infrastructure should be sufficiently strong to handle the international diffusion of all viable innovations. This would ensure that no resources were wasted.

Needless to say, this is hardly feasible. In an environment of technological change disequilibrium is the rule rather than the exception. Moreover, some elements of infrastructure are more amenable to government development than others, suggesting that the creation of temporary imbalance may itself be a means of strengthening the infrastructure.[5] By developing those general elements of infrastructure for which it is better suited, the public sector can create profit opportunities for private entrepreneurs to complete the missing elements. Nonetheless, any strategic plan for development of the infrastructure must strive to keep these imbalances, both spontaneous and contrived, within appropriate limits.

These limits derive, in the first instance, from the need to use a limited development budget in the most efficient possible way. They must also take into account, however, the very serious considerations that arise from the realities of international factor mobility.

In a free society that allows international movement of the factors of production, and especially of highly skilled labor, large imbalances in the infrastructure can create large disparities between the marginal product of these factors at home and abroad, generating strong incentives for movement of these factors of production across national borders. Clearly the direction of movement is crucial in these cases: a surplus of trained manpower can attract foreign investment or it can cause a "brain drain." Thus in creating such imbalances careful consideration must be given to the projected direction of the factor flows.

5. In this connection it is worthwhile to recall Hirschman's (1958) unbalanced-growth thesis: some imbalances induce compelling corrective (market) forces, while the effects of others are permissive.

## Direct Support for R & D Projects

Development of a sound infrastructure is a necessary condition for industrial innovation. But there appears to be general agreement—among theorists and practitioners alike—that the economic performance of innovative industries will also benefit from a program of direct support for specific R & D projects. The support of practitioners is evident from the wide currency that such programs have gained. Theoretical support is manifest in a long line of analyses of the positive externalities generated by innovative activity, from Nelson's (1959) and Arrow's (1961) early work to the more recent contribution of Spence (1984) along the lines indicated above.

Such analyses have focused primarily on the need for R & D subsidies and on their optimal level. We take this a step further, by bringing these well-recognized theoretical considerations to bear on the practical issues of designing a functional, efficient R & D project support system. These issues are captured in the following four questions:

What are the individual criteria that a project must meet to receive support?

Should these criteria be applied neutrally or should the system be targeted to specific industries?

Should firms or nonprofit institutions be preferred as the locus of industrial innovation (or neither)?

How far into the project life cycle should public support extend?

Before going on to consider each of these issues separately, we sketch a simple theoretical model of R & D support that highlights the practical difficulties inherent in designing a project support system. A fifth practical issue of substantial importance, the appropriate instruments that should be employed in supporting R & D projects (direct subsidy, loans, tax relief, government procurement, etc.), is left for future study.

### A simple model

Economic theory views subsidies as a means for modifying the economic behavior of firms and individuals so that it is in greater consonance with the general welfare of society. Thus a subsidy might be used to make profitable unprofitable ventures that are socially worthwhile, or to increase R & D outlays when less R & D is performed than is socially desirable. But there would appear to be little reason to subsidize R & D in projects that appear profitable, ex ante, without the aid of a subsidy, and where the scope of R & D is either optimal or unaffected by subsidies.

To fix ideas, consider a profit-maximizing firm evaluating a possible investment in R & D. Denote R & D outlays by $R$; and let $Q(R)$ and $B(R)$ denote, respectively, the expected net present value (ENPV) of private and social gains from R & D outlays gross of these outlays, discounted at the appropriate rates (fig. 9.2). Then the net private and social benefits are, respectively,

$$P(R) = Q(R) - R \qquad \text{and} \qquad S(R) = B(R) - R.$$

Denote the socially optimal level of R & D outlays by $R^*$; so that $B_R(R^*) = 1$, where the subscript denotes a derivative. Then the market achieves this result unaided only if $R^*$ is also the privately optimal level of R & D expenditures; i.e., only if the following necessary (but not sufficient) conditions hold:

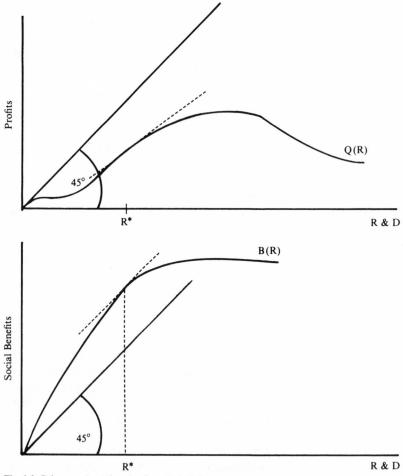

Fig. 9.2. Private and public gains from R & D investments.

(i)   $Q_R(R^*) = 1$,

(ii)   $Q(R^*) \geqq R^*$.

Thus the market may fail in the allocation of resources for innovation on either or both counts (as it does in fig. 9.2), in which case our optimal subsidy would comprise two parts: a pro rata subsidy, at a rate of $1 - Q_R(R^*)$ [if $Q_R(R^*) > 1$, this would be a tax]; and a lump sum subsidy of $[1 - Q_R(R^*)] R^* - Q(R^*)$. [The need for a lump sum subsidy arises only if $Q(R^*)/R^* < Q_R(R^*)$.]

Clearly such a model rests on many simplifying assumptions that restrict its application: it is static; it does not allow for competitive interaction; it ignores the restrictions of an inherently imperfect capital market; and it equates innovation with R & D. But the most severe limitations on its application stem from the information it requires, about both private profit expectations and social benefits.

With regard to the former, the difficulty stems not only from an objective lack of information (the firm has some notion of its expected profits) but also from the strong incentive the firm would have to belittle these profits in its dealings with the authorities, so as to gain as large a subsidy as possible.

As for the measurement of social benefits—those are far more nebulous. An individual R & D project may be insignificant in itself and yet be an important link in a developing chain of innovations. Indeed, Nelson and Winter have consistently made the point that one of the important functions of R & D subsidies is to encourage experimentation and preliminary work leading to a richer set of feasible R & D projects (Nelson and Winter 1982). Obviously, in placing a value on the marginal benefits from future research projects (yet unknown!), one cannot pretend to any degree of precision.

A second difficulty that arises in assessing prospective R & D projects singly derives from possible competitive interaction between innovations aimed at overlapping (or even identical) markets. In such a case innovation generates negative externalities, and as the theoretical work of Barzel (1968) and Hirschleifer (1971) indicates, this may give rise to excessive innovation.

Finally, a third difficulty stems from the stochastic dependence between prospective R & D projects. Financial theory teaches us that when such dependence exists, an optimal portfolio of risky assets will take it into account, seeking to balance the risk so as to eliminate as much of it as possible.[6] This complex problem, i.e., the optimal composition of a national portfolio of

6. The portion of risk that can be eliminated in this way is generally referred to as "unsystematic risk" in the financial literature.

R & D projects, has not yet been dealt with analytically, to the best of our knowledge.

All of the above suggests that the allocation of public support for R & D projects cannot be determined, practically, with full adherence to theoretical rigor. Instead we seek practical indications of conditions that are likely to justify public support. Specifically, we would like to characterize the type of R & D project for which private profits do not justify investment at any scale $[Q(R) < R$ for all $R]$ but for which social benefits outweigh costs at some level of R & D outlays $[B(R) > R$ for some $R]$. These are likely to be projects with a large disparity between private and public gains.

### Individual criteria for project approval

Turning now to the question of criteria for project approval, we note that in practice technological criteria for approval are increasingly giving way to commercial criteria. For the reasons presented in the previous paragraphs we view this as a positive development but argue that it should be taken a step further: commercial criteria should be replaced by economic criteria along the lines set forth above. Discrepancies between private motives and public goals provide the proper basis for administering R & D support policies.

On a more specific level we suggest that there are three sets of circumstances that may justify R & D project support. Chronologically, the opportunity may first arise in the early stages of industry development, when R & D subsidies may be employed to foster the accumulation of general experience, as an element of infrastructure. This issue was examined above and will not be taken up again here. Beyond that, certain industries may offer an opportunity to exploit market imperfections by subsidizing R & D. Strategic considerations of this type were discussed briefly above and will be examined further in the following subsection. Finally, the general circumstances that limit the gains from innovation by private investors, also outlined above, provide the third possible basis for R & D subsidies. This is the topic of the present subsection.

That the generation of externalities is an appropriate basis for subsidization is a well-established economic principle familiar from a variety of applications. The point we want to make here is that in the context of research subsidies in an open economy the incidence of externalities generated by the innovative process must be considered. It matters very much who will produce the good that results from the innovation and who will benefit from it. And it matters who owns assets that the innovation enhances and who owns those that it renders less valuable. Two polar examples should serve to illustrate this point.

Consider, at one extreme, a small Israeli subsidiary of a multinational corporation developing integrated circuits which are then manufactured in the Far East and sold to an American automobile manufacturer. Business prospects may be excellent, but these will be accounted for in investors' decisions in any event. Local externalities, however, seem slight, limited perhaps to those stemming from the labor requirements of the R & D effort. Interfirm spillovers are also likely to be minimal in the absence of a local automobile industry.

At the other extreme, the development by Israeli innovators of irrigation equipment for arid zones has generated external effects at several levels. There is extensive local manufacture of the equipment by a well-developed domestic plastics industry; there is widespread use of the innovation by local agriculture; and the produce is at least partly consumed by the domestic market. Moreover, the small scale at which production becomes efficient and the relative ease of imitation make for a competitive industry in which spillovers are hard to prevent, and the appropriation of benefits by innovators is consequently limited.

Indeed, taking this line of reasoning a step further, there may well be some research activities which should be discouraged, because of the adverse external effects they are likely to have on important national assets. This might apply, for example, to innovative research in the field of nuclear power generation in a country with a large oil surplus. It is widely recognized that the sale of existing technologies though profitable for the individual firm may well be detrimental to the national economy, and the same reasoning applies to the generation of new technologies.

Moreover, while the magnitude of the externalities that a research program is likely to generate cannot generally be predicted to any reasonable degree of precision, the incidence of these externalities may well be quite clear even at the very earliest stages of the program. This would then provide a useful criterion for discriminating between research programs competing for public subsidies. Projects likely to produce benefits for domestic consumers through domestic manufacture would rate the highest priority, while those geared primarily for licensing to foreign manufacturers would rate the lowest. Projects of the latter kind may well be of great benefit to the country, but this is likely to be so only if they are also profitable to the private investor, in which case subsidy is unnecessary.

The intermediate case, of goods produced domestically for export, may also offer substantial externalities. Jobs are created, often with higher than average wages. And foreign currency is earned, improving the balance of payments. In a country with a large trade deficit and, concomitantly, a large foreign debt, this has the beneficial external effect of improving the terms on which it receives credit, and thus strengthening its currency. This is a general argument for supporting export-oriented industries. There is an additional argument that applies more specifically to innovative industries that compete in world markets. This is taken up in the next subsection.

Neutrality

We use the term neutrality, here, to refer to the uniform application of universal criteria to all projects applying for R & D subsidies, irregardless of the industry, technological area, or product class to which they belong. In early stages of economic development, and generally in the absence of reliable information on the commercial prospects and externalities likely to be associated with different projects, a neutral subsidy policy will take the form of a flat rate of subsidy to all projects meeting some minimal criteria of competence and integrity.

In later stages such policy will be modified in accordance with information gathered regarding the relevant economic variables. This might induce a shift from the formal neutrality of the early stages without implying abandonment of the principle of neutrality. Thus, given the costs of a thorough examination of all projects submitted, the use of signals or rules of thumb may well be justified on economic grounds, at least as initial screening devices. These rules of thumb might be aimed at capitalizing on the strengths of the country's industrial infrastructure; enhancing the value of its natural resources; providing a productive channel for underemployed human capital; etc. Their formulation might naturally rest on a classification of projects by industry, technology, or product class, inter alia, without there being a departure in principle from a policy of neutrality.

A policy of neutrality in the broad sense defined above is justified, in our view, when government intervention in R & D is aimed at correcting for externalities. Israel's successful experience with a neutral policy of research subsidies would seem to bear this view out.

Departure from such a policy might be dictated, however, by large lumpy investments that are likely to significantly strain the country's limited resources. Such investment decisions are inherently nonneutral and must be judged in a wide context that can take into account their strategic implications.

Obviously, the strategic dimensions of large R & D investments are most commonly felt in small economies. For example, the development of modern fighter aircraft in countries such as Israel and Sweden drains vital engineering skills from other sectors of the economy. They are also sorely felt in developing countries undertaking large, high-technology projects such as the nuclear programs in India, Pakistan, and Iran. But they can also be felt in large, developed countries, as in the case of the United States space program in the 1960s.[7]

---

7. It does not strike us as coincidental that R & D investments with national strategic dimensions are often defense related. The obvious dangers inherent in such investments, which put a disproportionate number of eggs in the same basket, are more easily accepted when they can be justified by national security interests.

Finally, a deliberate departure from neutrality might be justified by strategic considerations of the type outlined in the second section, irrespective of project size. These would target a specific, imperfectly competitive international market still in its formative stages for national development and possible dominance through early subsidy of industrial R & D. Such subsidies would clearly be nonneutral, in any sense.

This form of targeted policy bears some resemblance to the type of industrial research policy commonly referred to as "picking winners." We want to stress, however, that it is much more stringent in its requirements. It stipulates a market in its formative stages that when mature will have room for only a small number of national industries. And the location of production facilities must follow the location of R & D activities. Moreover, the possible supranormal profits that may be gained must be weighed against the risks inherent in a bureaucratic decision involving an assessment of the future prospects of an industry in its early stages of development.

### The locus of industrial R & D

In practice, support systems for industrial R & D projects have been directed to both firm-based research and nonprofit institutions. Our detailed knowledge of Israeli experience with both types of system (see Teubal 1984) and general acquaintance with the experience of other countries indicate that while nonprofit research is most suitable for many projects in agriculture, health, and defense, the bulk of industrial research is best served by a system geared to supporting research in firms. This stems primarily from the responsiveness of firm-based research to the prospects of commercial success. These prospects generally bear a closer affinity to economic welfare than do the technological criteria to which nonprofit research organizations are more strongly oriented. Furthermore, directing R & D project subsidies to the individual firm creates indirect demand pressures for the other necessary ingredients of successful economic development of high-technology industries, viz., advanced manufacturing facilities, marketing capabilities, and engineering and managerial skills. The innovative firm receiving project subsidies thus serves as a crucial feedback mechanism in the complex process of adjusting and balancing the entire spectrum of direct and indirect support policies. Moreover, it must be noted that while the biases of the individual firm in allocating resources for R & D have received a preponderance of the attention of economic theorists, such biases must surely also exist in nonprofit institutions. That we know less about the biases inherent in these institutions is not a recommendation in their favor.

However, the above-mentioned advantages notwithstanding, there appear to be at least three fields of endeavor where a substantial proportion of R & D project subsidies must be directed to nonprofit institutes. These are agriculture, health, and to a lesser extent defense. All share a rich potential for scientific discovery, and, in addition, each appears to have its own specific reasons for the large role it plays in nonprofit research.

In agriculture it would appear to be a combination of a highly fragmented market structure and the inherent difficulties in appropriating the gains from many of the specific innovations arising from agricultural R & D. In health it is the unique nature of the product and the concomitant traditional dominance of nonprofit institutions in conducting medical research. And in defense it would appear to be a combination of the final product being a public good, the need to be constantly on the forefront of scientific knowledge, and again the special intensity of demand for the final product. Indeed, in the last two cases there is a clear divergence between business considerations and economic welfare which induces a preference for a good deal of non-profit-oriented research.

Finally, past experience indicates that the natural commercial orientation of the individual innovating firm is likely to be reinforced if the institution administering R & D support is also economically or commercially oriented. Thus a policy of R & D support is more likely to follow commercial or economic principles if it is under the auspices of a ministry of industry. Conversely, its administration by a ministry of science and technology is likely to give greater weight to scientific and technological merit.

### The scope of project support

R & D is only one stage in the innovation process. Manufacture and marketing are subsequent phases that are also essential for successful application and diffusion of innovations. This raises the question of whether public support for innovation should extend beyond the development stage. Such support is generally not extended in practice; however, there are arguments pro and con, and these should be weighed in the context of specific projects.

The main argument for extending support beyond the R & D phase is that supporting R & D activities alone creates an imbalance in the system. A surfeit of inventions is generated, many of which are consequently not implemented. This argument can be interpreted in two different ways. The first is that the manufacturing or marketing infrastructure cannot assimilate all the inventions that emanate from the innovating sector. This was considered in the previous section, and we will not repeat the discussion here except to say that such circumstances indicate a need for development of the relevant elements of the infrastructure or realignment of the R & D support system so that it generates innovations that match the strengths of the infrastructure. Support for the manufacturing and marketing stages of technologically successful R & D projects may or may not be the economically efficient way of achieving this.

A second type of imbalance occurs at the firm level between R & D outlays and expenditures on production and marketing whose objective is implementation or commercialization of R & D results. Here we must distinguish the case where external benefits to society not capturable by the private firm are generated at the invention stage (e.g., spillovers of technological knowledge) from the case where the externalities result from commercialization (e.g., externalities from the "innovation" derived from additions to user surplus).

In the former case, preferential subsidization of R & D vis-à-vis implementation is justified in principle. Moreover, this conclusion is probably reinforced by the presence of capital market imperfections, which severely limit the market's ability to spread the risks of R & D. In the latter case, the target of government support should be the innovation process as a whole and not only R & D, or any other stage. Earmarking the subsidy for R & D alone will produce a distortion in that the innovating firm will tend to replace high-cost marketing and production effort by low-cost (but socially dear) R & D activities. Thus when the externalities from innovation are a result of commercialization, there seems to be no reason to alter the overall balance attained by market forces between R & D and implementation. This may require, however, close coordination between the R & D project support system and other preexisting support schemes that bear on the innovation process.[8]

Finally, the case has been made that the manufacture and marketing of innovations merit public support because innovators exhibit an inherent bias against such activities, being more technologically oriented. Such support would then arouse their awareness. This is an argument which must be applied carefully. Awareness can be aroused through the dissemination of information by extension services of the kind successfully employed in promoting the diffusion of agricultural expertise. Widespread public support for large-scale marketing efforts that cannot find a commercial sponsor seems an expensive way of going about this.

### Summary

In this paper we develop a conceptual framework for the formulation and assessment of technology policy. This framework is a synthesis of the neoclassical market failure approach to normative economics and the Schumpeterian view of innovation as a process of development. It is presented in the context of the concrete policy issues which it must address.

We distinguish between two principal components: development of an infra-

---

8. However, in an evolutionary context, preferential support for R & D (or other stages of the innovation process) may be necessary to compensate for the cumulative effects of past distortions.

structure for innovation and support for specific R & D projects. We view the former element as a strategic activity with wide ramifications, requiring coordination at the highest national level. The latter provides tactical support and can therefore be managed, within broad guidelines, at the bureaucratic level.

We define the innovation infrastructure as a combination of elements that support the innovation process while exhibiting economies of scale that transcend the needs of the individual firm. These elements fall into four principal categories: research and development, finance, manufacture, and marketing. We then go on to describe the role of each component in the innovation process and the reasons why public intervention may be warranted to ensure that this role is performed adequately. Such intervention must promote the availability of the necessary skills and capabilities, organizational structures, and physical facilities needed at each stage of the innovation process and which the market will not find it profitable to provide. Moreover, it must carefully control any imbalances in the infrastructure, both to avoid wasting resources and to prevent the outflow of valuable factors of production.

The other key element of innovation policy, a support system for specific R & D projects, is molded in large part by the answers to four key questions considered in this study. They are: What are the individual criteria that a project must meet to receive support? Should they be applied neutrally across industries? Where is the preferred locus of industrial innovation? Which stages of the innovation process should be supported? Our detailed answers to these questions stress the advantage of economic criteria for project approval over technological and commercial criteria. These focus on the domestic incidence of positive externalities as the main justification for supporting specific projects, and they suggest a policy of neutrality, in principle, in the administration of project support systems.

Such systems, however, may also have strategic dimensions. In the early stages of market development project support systems have an important role to play in the building up of basic design capabilities and other relevant skills. In this respect project support aids in strengthening the infrastructure. Beyond this, strategic support for targeted industries may be warranted under specific conditions of imperfect competition in an international market still in its formative stages.

Finally, our work identifies several important issues for future study. Chief among these are a better analytical perspective of the dynamics of infrastructure development; a quantitative approach to determining the optimal size of the various budget elements supporting industrial innovation; and a comparative analysis of the various instruments of innovation policy (subsidies, loans, procurement, patent protection, etc.).

### References

Arrow, K. 1961. "Economic Welfare and the Allocation of Resources for Invention." In *The Rate and Direction of Inventive Activity: Economic and Social Factors,* edited by R. Nelson, 609–25. Princeton: Princeton University Press.

Barzel, Y. 1968. "Optimal Timing of Innovations." *Review of Economics and Statistics* 50:348–55.

Brander, J. A., and B. J. Spencer. 1985. "Export Subsidies and International Market Share Rivalry." *Journal of International Economics* 18:83–100.

Hirschleifer, J. 1971. "The Private and Social Value of Information and the Reward to Inventive Activity." *American Economic Review* 61:561–74.

Hirschman, A. 1958. *The Strategy of Economic Development.* New Haven: Yale University Press.

Justman, M. 1984. "The Welfare Economics of Research Policy in an Open Economy." Working Paper. Ben-Gurion University.

Magaziner, I., and T. Hout. 1980. *Japanese Industrial Policy.* London: Policy Studies Institute.

Nelson, R. 1959. "The Simple Economics of Basic Scientific Research." *Journal of Political Economy* 67:297–306.

Nelson, R. 1983. "Government Support of Technical Progress: Lessons from History." *Journal of Policy Analysis and Management* 2:499–514.

Nelson, R. 1984. "Policies in Support of High Technology Industries." Working Paper no. 1011 (revised). Institution for Social and Policy Studies, Yale University.

Nelson, R., M. Peck, and E. Kalacheck. 1967. *Technology, Economic Growth and Public Policy.* Washington, D.C.: Brookings Institution.

Nelson, R., and S. Winter. 1982. *An Evolutionary Theory of Economic Change.* Cambridge: Harvard University Press.

Peck, M. 1983. "Government Coordination of R & D in the Japanese Electronics Industry." Yale University. Mimeo.

Rao, K. N., and C. Weiss. 1982. "Government Promotion of Industrial Innovation." World Bank. Typescript.

Rothwell, R., and W. Zegveld. 1981. *Industrial Innovation and Public Policy.* London: Frances Printer.

Spence, M. 1976. "Product Selection, Fixed Costs and Monopolistic Competition." *Review of Economic Studies* 43:217–35.

Spence, M. 1984. "Cost Reduction, Competition and Industry Performance." *Econometrica* 52:101–21.

Tassey, G. 1982. "Infratechnologies and the Role of Government." *Technological Forecasting and Social Change* 21:163–80.

Teubal, M. 1982. "The R & D Performance through Time of Young, High-Technology Firms: Methodology and an Illustration." *Research Policy* 11:333–46. (Chap. 6 in this volume.)

Teubal, M. 1984. "Neutrality in Science Policy: The Promotion of Sophisticated Industrial Technology in Israel." *Minerva* 21 (Summer): 172–97. (Chap. 8 in this volume.)

Urban, P. 1983. "Theoretical Justifications for Industrial Policy." In *Industrial Policies for Growth and Competitiveness*, edited by F. G. Adams and L. Klein, chap. 3. Lexington, Mass.: Lexington Books.

Westphal, L., L. Kim, and C. Dahlman. 1984. "Reflections on Korea's Acquisition of Technological Capability," Washington, D.C.: World Bank.

Williams, R. 1983. "Technology Policy." Typescript.

We appreciate comments on a previous draft from R. Nelson, K. Pavitt, and two anonymous referees and from participants in seminars given at the Science Policy Research Unit, University of Sussex, and at the Instituto Torcuato Di Tella and the Fundación Mediterranea, Argentina.

# 10 Government Policy, Innovation, and Economic Growth
## A Study of Satellite Communications

Morris Teubal
and Edward Steinmueller

### Introduction

The path leading to the maturity of industrial sectors is generally associated with a deceleration in their rates of growth. This was pointed out almost 50 years ago by Simon Kuznets. An important implication of this trend is that any economy wishing to maintain its *aggregate* growth rate must generate a continuous flow of new industries (Kuznets 1930; Burns 1934). Federal R & D may stimulate this process by inducing the private sector to undertake radical innovations which would otherwise not be undertaken or whose introduction would be seriously delayed. The reduction in the technological and market risk of the innovations, consequent to federal R & D programs, may counteract the forces responsible for deceleration, including the observed tendency of some established industries to follow an evolutionary rather than a revolutionary path of technological development.

Our purpose here is to illustrate and give some structure to the above in the context of and based on the experience with satellite communications. In addition, we will attempt to measure some aspects of the social benefits derived from this new industry and to address the question of the share of these benefits which may be attributable to NASA. The report is organized in three sections. The first is a historical description of technological developments in communications satellites. Its emphasis is on the efforts which culminated in the first commercialization of the technology in 1965. An attempt is made to describe the events in the context of our knowledge of innovation processes in general and to describe the activities and impacts of NASA. The next section addresses the issue of the benefits to society from NASA's efforts in satellite communications. A distinction is proposed between a "baseline" communication satellite technology and subsequent developments. An argument is proposed attributing all the economic benefits of the former to NASA. A computation of part of these benefits is also presented. Finally, we discuss some of the issues which emerge from this case study.

## A Historical and Technological Survey of the Development of Communications Satellites

The development of communication satellite technology and NASA's involvement in the area can be broadly divided into three stages: invention, demonstration of feasibility, and commercial development. The invention stage includes a series of critical experiments which determined that no physical barriers to orbital transmission and reception of signals existed. In the demonstration of feasibility stage a wide variety of system designs were tested with important knowledge and skill-creating implications for emerging supply companies such as Hughes Aircraft and General Telephone and Electronics. Finally, with the development of the quasi-private communications companies COMSAT and INTELSAT, commercial users undertook the contracting of commercial satellite communications systems and a variety of ground station installations, a "reduction to practice" of the new technology. NASA played an active role throughout its period of involvement (1958 till the early 1970s) in the development of the communication satellite technology by providing a research and development infrastructure and by contracting for operating systems which steadily increased in capabilities and performance. NASA's research and development decision-making process involved a series of committees and solicited proposals which sought to define the major technological problems at each stage of the development process and to plan a series of design research tasks which would solve these problems. This infrastructure also was a basis for determining the nature of contract specifications when specific systems were developed, thus coordinating technological policy with implementation.

Communications satellites were visualized in the 1950s as an alternative technology for coping with a basic constraint of terrestrial telecommunications, the curvature of the Earth. The Earth's curvature is a barrier because the propagation of electromagnetic radiation is linear. Before 1960 the available solutions were direct connection (undersea cables or a series of "links" on land) or high-frequency radio transmission which utilized the reflectivity of the ionosphere to "bounce" signals over the horizon. The latter phenomenon of ionospheric reflectivity caused concern about the viability of satellite signal transmission because of the uncertainty of how efficiently signals could be sent through upper-atmosphere layers. The SCORE launch demonstrated that signals could indeed be transmitted and received with reasonable efficiency from an orbital position (Pritchard 1977). The next step in experimenting with the idea of satellite communications was the development of a passive reflector which would serve the same purpose as ionospheric reflection with more efficiency and reliability. The problems of efficiency and reliability had limited the growth of intercontinental radio communications.

### The *ECHO* program

The *ECHO* program was based on this principle of a "passive" communications satellite. *ECHO* did provide a reliable means of "bouncing" signals over the horizon. However, *ECHO* by no means solved the problem of efficiency. *ECHO* proved that the passive satellite approach led to a drastic limitation of power available to the receiving station because of the dispersion of transmitted and reflected signals. In the *ECHO* experiment only one part in $10^{18}$ of the transmitted power (10 kW) was available to the receiving antenna (Pritchard 1977). Of course this limitation was known before *ECHO*'s launch, and the development of "active" satellite systems was presumed to be the only really viable means of achieving substantial communications capability. The "active" communications satellite receives a signal and retransmits an amplified facsimile of the signal (usually with a frequency shift) (Pritchard 1977). The problems of active satellite design are analagous to designing telecommunications links for terrestrial microwave systems with the important added problems for the space environment, weight constraints, power limitations, and impossibility of maintenance. These problems were the fundamental elements of NASA's demonstration of a feasible satellite communications technology and determined the design problems of the first of NASA's major satellite programs. Work on active satellite design could not proceed until NASA was granted authority in 1960 to work in formerly Defense Department technological territory by revising the 1958 DOD/NASA agreement (Midwest Research Institute 1971). This agreement had to be further abridged in 1961 to allow NASA to engage in geosynchronous satellite research (Midwest Research Institute 1971). Defense interests may have prohibited private development of communications satellites because of the difficulty of controlling access to knowledge of such privately developed systems. NASA, by pioneering the technology, demonstrated the commercial potentials, and this led to pressure by a great many parties to allow commercial development. What would have occurred without NASA is of course counterfactual speculation, but it is probable that NASA/DOD relations bridged the barrier to commercial development of communications satellites.

### The *RELAY* program

The *RELAY* program successfully pioneered a series of major innovations including one of the first uses of traveling-wave tube (TWT) amplifiers in space, a number of important firsts in types of data transmitted, and ground station technique improvements (Pickard 1963, 1965). The *RELAY I* mission was designed to provide a great deal of engineering data allowing the optimization of communication and other subsystem design as well as enhancing understanding of the radiation (high-energy particle) space environment (Pickard 1963, 111, 115; 1965, 681–87). Technical characteristics of the satellite transmitter were chosen to conform with existing international standards (Pickard 1963, 116, 118). In spite of serious problems with the power system and a last minute discovery which necessitated that the TWT as designed operate in a pressurized container, *RELAY I* did accomplish its experimental goals. The mass of data accumulated by this program should be stressed. It was possible to assess the accuracy of rebroadcasting from space by determining the exact deviations of received from transmitted signals (Pickard 1963, 113). Subsystems such as power, elementary attitude control, telemetry, and temperature regulation were demonstrated as workable, and important modifications (in power and TWT design) were made for subsequent satellites. The basic ground station problems of tracking and data acquisition were also solved for the first generation of communications satellites.

The importance of *RELAY I* to the development of communications satellites is that a basic design solution was successfully implemented paving the way for incremental improvement and commercial adaptation of subsystem designs. Once a design configuration is achieved which will perform in a satisfactory manner, efforts can be made to enhance its performance by fine-tuning or partial redesign. A successful basic system offers a base for add-on systems with different capabilities, so subsequent improvements can be made with relatively low risk. A related strategic question is the choice of technologies with promising prospects for incremental improvement. It is apparent that certain technologies, such as the *ECHO* passive concept, have stringent and predictable limits for possible improvement. In other cases, such as solar cell technology, limits to improvement seriously constrain short-term improvements but occasional major breakthroughs may dramatically enhance performance.

A second important lesson of the *RELAY I* program is the parallel it provides to NACA and NASA aeronautic activities such as those undertaken since 1940 (Hartman 1970). These efforts, by accumulating a large data base in an area of technological knowledge, have permitted a variety of military and civilian engineering designs to be created at less cost and with greater possibility of success. This type of knowledge accumulation also has important implications for the competitive structure of industrial organization because of NASA's disclosure requirements, which allow any firm access to the technological data base. Economically, this is the proper treatment of knowledge in the sense of maximizing public benefit (Arrow 1961). Information should be made available at a price equal to its transmission or publishing cost. The competitive implications of this system of knowledge accumulation arise because the costs of entry in an industry are in part based upon the costs of accumulating the knowledge base necessary to produce a competitive product (Mueller and Tilton 1969). Further investigation of NASA promotion of competition in high-technology industries would be an important adjunct to studies of benefits of NASA programs.

### The *SYNCOM* program

One critical limitation of the nonsynchronous satellite design of lower power, such as *RELAY I*, was the necessity for tracking (movable) antennas. This necessarily limited the number of facilities which could reliably receive extended communications because of the high cost of each station. It also dictated a very high investment in satellite systems if continuous communications were to be an objective, since it implied a string of satellites so that at least one satellite would always be above the relevant receiving area. For these reasons, the possibility of a major redesign of the basic satellite–ground station system was proposed by Dr. H. Rosen of Hughes (Rosen 1963). Hughes was unable to sell this idea to the Department of Defense or any corporation large enough to support the development costs beyond the exploratory stage accomplished at Hughes.[1] We may speculate that the major reason for the difficulty of selling the Hughes system was the risk of an entirely new system when a known, albeit expensive, solution was known to exist. It is also important that J. Pierce, of Bell Laboratories, expressed severe criticisms of the proposed geosynchronous satellite because its altitude would entail a delay in communications of 0.46 seconds, which Pierce believed detrimental to "quality" of telecommunication service. However, NASA finally was able to support research in the area, which led to the *SYNCOM I*. In order to do this NASA had to obtain another agreement with DOD concerning geosynchronous satellites, an area which had previously been the exclusive domain of DOD by agreement with NASA. Authorization was achieved in 1961 permitting NASA and Hughes to jointly develop the *SYNCOM I* satellite (Midwest Research Institute 1971, 12).

The fundamental design problems of achieving a synchronous orbit and broadcasting a continuous signal to Earth were solved by Rosen and his research group at Hughes. NASA awarded a contract to Hughes in August 1961 leading to the February 1963 launch of *SYNCOM I* (Midwest Research Insti-

---

1. Interview with Dr. H. Rosen, July 1978.

tute 1971, 14). Unfortunately, a nitrogen tank explosion destroyed the communications capability during launch. The basic problem of orbital placement was solved in spite of this failure: ground station tracking revealed *SYNCOM I* had a nearly ideal orbit. The failure of *SYNCOM I* to serve its communications purposes, in spite of an expenditure of $20 million to develop, manufacture, and launch the satellite, would have been a severe blow to the development of an exclusively private satellite industry. Because such risks were known by NASA and direct payback was not an issue in the calculus of mission decision, NASA prepared and launched the backup satellite *SYNCOM II* in July of 1963 (Midwest Research Institute 1971). *SYNCOM II* was not designed to be a truly geostationary satellite but rather to maintain an orbit which caused it to move a distance of 33 degrees north and south of and perpendicular to the equator (Midwest Research Institute 1971, 16).

Midwest Research Institute has cataloged a long list of improvements incorporated in the *SYNCOM II* program including technologies for transfer orbit, antenna array, automatic station keeping, and a variety of improvements in power and communications subsystems (Midwest Research Institute 1971, 15–16). The major experimental purpose of *SYNCOM II* was to establish the feasibility of the semistationary orbit and provide engineering data on communications design. *SYNCOM II* was so successful in establishing the viability of synchronous satellites that civilian communications satellite systems have subsequently uniformly been of the synchronous design. Back in 1961, the viability of a synchronous orbit solution was severely doubted. The *ADVENT* program, an Army synchronous satellite system, had been running into difficulties, which eventually led to its demise. Thus, if *SYNCOM* had not been successful, there is every reason to believe that the present predominant technology for communications satellite orbits, synchronous orbits, would have emerged at a much later date if at all.[2] This success was the basis for taking much of the risk out of developing commercial communications satellites. The *SYNCOM III* satellite, which achieved a near-perfect geostationary orbit in August of 1964, was followed less than one year later by *INTELSAT I*, the first commercial communications satellite.[3]

The *SYNCOM III* achieved a truly geostationary orbit and demonstrated the technology of automatic station keeping, which was another link in the development of synchronous satellites. Two aspects of this station-keeping capability were important to further developments, the discovery that ground

2. Interview with Dr. H. Rosen, July 1978.
3. *TELSTAR* preceded *INTELSAT I* but was only in limited use and not available to the general market.

tracking was inferior in accuracy to in-orbit station keeping and the discovery of the Earth's "tri-axiality," pear-shapedness, which meant the impulse power to preserve stationarity was significantly less than originally anticipated. Contrary to expectations, onboard position sensing was much more accurate than terrestrial observation. The unexpected accuracy of onboard station keeping, position sensing, and correction set a standard of accuracy which would for several years focus technical efforts on improving ground station tracking. This phenomenon of focusing is also an important aspect of technological change (Rosenberg 1976). In economic terms the nature of focusing mechanisms is that they provide a clear standard by which to measure quality change. The potential user can use standards and specifications to relate the cost of using a particular system to its capabilities. The latter development, the discovery of the pear-shapedness of Earth (though the analogy is more appropriately an overripe peach that has been picked up firmly in the Northern Hemisphere), is an example of a rare but significant happening in engineering, the discovery of a new physical principle while engaged in solving problems within a well-defined model. While much engineering work does not benefit from a scientific model (the phenomena are too complex to be completely specified or too new for theory to have caught up with practice), it is somewhat rare for engineering to force modifications in the models it does use. In this case the technology of geosynchronous satellites was a more attractive option when the knowledge about the physical state of the world was changed by empirical observation. By 1964 the standards and specification of a particular type of communications satellite were set, and the major questions facing the commercial developers were the costs they would face in implementing the technology rather than developing it.

The pattern of incremental improvement noted earlier was apparent in the *SYNCOM* program. The development of a nonpressurized traveling-wave tube amplifier as well as changes aimed at eliminating the specific catastrophe of *SYNCOM I* made *SYNCOM II* an improved satellite. A series of incremental changes occurred with *SYNCOM III*; solar power collectors delivered 32.5 W versus 28 in *SYNCOM II*, bandwidth went from 0.5 mHz to 10 mHz, and a TV channel was added (Midwest Research Institute 1971, 15). By 1963 the potential of satellite communication was well established, and NASA began a series of advanced project plans including a fourth, "advanced" *SYNCOM*, in case of a *SYNCOM III* failure (Midwest Research Institute 1971, 20). Basically, the advanced *SYNCOM* was a further demonstration of technical feasibility of certain systems as well as a model incorporating incremental improvements in a number of subsystems (power and number of channels in particular). Point-to-point transmission problems and possibilities were well understood by 1964. NASA was authorized in February 1964 to expand the potentialities of satellite communication in a new program entitled Applications Technology Satellites. *ATS* represents the beginning of the "reduction to practice" stage in satellite development. A consideration of the technological planning process leading to the *ATS* program is important in understanding how NASA successfully bridged the gap between doing work which was in many respects purely a demonstration of feasibility and doing work which had a larger development component.

### The *ATS* program

The *ATS* program proposal grew out of planning for the advanced *SYNCOM* and a study to define technological needs which began at Goddard Space Flight Center in November 1963. A whole series of possible experiments was outlined and missions were built up around them. The resulting *ATS* program was to explore a broad range of possible uses of satellite communications. After the program was approved, further experiments were solicited from outside researchers. NASA attempted to outline broadly the extent of the new technology, providing relevant engineering data to potential commercial and other government entities. NASA also proposed to develop a series of radically new ideas, that is, to underwrite risks across a very broad range of innovations. In proposing such a course of action it was essential that NASA confine its attention for the most part to issues of technology, not directly addressing the commercial prospects or cost-effectiveness of a particular innovation. With the clarity of hindsight, this appears to have been an excellent choice, for costs of many subsystems have come down and capabilities have increased dramatically beyond expectations, particularly in the area of electronics technology, which forms the basis for much of the communication satellites' operating systems.[4]

The contributions of the *ATS* program to technological progress in satellite communications are numerous and distributed over a considerable number of speciality areas. Table 10.1 itemizes some of these contributions. The specific intent of the *ATS* program was to explore new technologies for application to later satellites. Among these technologies were capabilities for multiple access,

4. During the course of the present study the authors examined several other NASA programs, including automated flight surface control and the SETI program. In each, the development of a planning mechanism centered on an identification of technological issues, and the needed experimentation to clarify or resolve these issues was the operative focus of the planning mechanism. We might suggest, to use economic parlance, that this is to NASA's comparative advantage and therefore is the best course of action.

Table 10.1. Mission objectives of *ATS I* through *ATS V*

| Designation | Launch date | Primary mission objectives |
|---|---|---|
| ATS I | 7 Dec. 1966 | Multiple access; phased array antenna; aircraft communications; spin-scan camera; station keeping; small terminals. |
| ATS II | 5 Apr. 1967 | Gravity-gradient stabilization; SHF multiple access frequency translation aspect sensors; active dampers. |
| ATS III | 5 Nov. 1967 | Mechanically despun antenna; color spin-scan camera; OPLE; APT; aircraft communications; range and range-rate. |
| ATS IV | 10 Aug. 1968 | Gravity-gradient; stabilization; cesium ion-engine thrustor; microwave SHF and UHF communications. |
| ATS V | 12 Aug. 1979 | ATC on L-band; CG stabilization; active and passive dampers; multiple spiral antenna arrays; heat pipes; 10 and 31 GHz millimeter-wave propagation; multiple access C-band. |

*Source*: Midwest Research Institute 1971, 23.

aircraft communication, spin-scan cameras (color and black and white), position sensing, operation in new frequency bands, and improvements in information coding technique (Midwest Research Institute 1971, 23). The 1974 *ATS VI* satellite was an important breakthrough in achieving high orbital broadcast power and dramatically reducing the size and cost of Earth-based receivers (Redisch 1975).

## The transfer of NASA-sponsored technology

The most important observation about these technical improvements is that some of them "pace" the developments in commercial satellite technology. That is, NASA programs generally led in the development of the major technological solutions to commercial satellite problems. In table 10.2 we have compared a series of such major design solutions transferred from NASA contract to commercial satellite systems (often constructed by the same contractors) (Midwest Research Institute 1971, 47–48). The importance of NASA's technological lead is twofold. First, NASA had a significant lead in experience over the commercial sector and was pursuing objectives of a more technological character (thus NASA was perhaps less constrained by cost considerations). Second, NASA appears to have adopted an explicit policy of taking relatively large risks in order to extend technical capability, whereas user firms such as COMSAT have been more cautious in their explorations of new technology, tending to make improvements within existing system design. The latter point is particularly relevant for current NASA policy.

Table 10.2. NASA's program contribution to commercial satellite technology

| Type of service | Total system | | Space segment | | | | | | Earth segment | | | |
|---|---|---|---|---|---|---|---|---|---|---|---|---|
| | System capability (TV, voice) | Operating characteristics | Orbit, stabilization, station-keeping | Satellite power | Satellite antenna | Satellite transponder | Channel capacity | Down-link frequency | Ground-station transmit antenna | Up-link frequency | Ground-station receive antenna | Ground-station receiver |
| International satellite system | SYNCOM first synchronous trans-oceanic TV | advanced SYNCOM II, ATS I (multiple access) SR&T (coding and modulation) (PCM-FM) | SYNCOM II (altitude sensors) ATS I (spin stabilization) (pulse jet control) (polang attitude) | ATS I (low-energy proton damage) | ATS I (electronic phased array) ATS III (mechanically despun array) | SYNCOM III (broadband 25 MHz) ATS I ATS III (dual mode) | SYNCOM II and III ATS I (building block) (channelization) | SYNCOM II and III ATS V (millimeter waves) | (Cassegrain feeds) ATS I ATS III (pseudo-monopulse) SR&T | SYNCOM II | SYNCOM II (margins) ATS I SR&T | SYNCOM II ATS I SR&T (threshold demodulators) (liquid $N_2$ cooled paramps) (wideband 1-F) |
| Broadcast distribution (point to multipoint) | | SR&T (Multi-path protection) | SYNCOM III (tri-axiality and gravity potential) | SR&T (RF power handling) (RFI problems) (heat pipe cooling) | ATS III SR&T (shaped area coverage) | SR&T (high-output TWTs) | | | ATS III (12 kW multi-cavity klystrons) | SYNCOM III | ATS I ATS III ATS F (small stations) | ATS III ATS V (paramps 30 GHz) |

| | | | | | | | | | | | | |
|---|---|---|---|---|---|---|---|---|---|---|---|---|
| Direct broadcast satellites (point to wide area) | *ATS F* | *SR&T* (digital multiple access) | *ATS F* and *G* (low-thrust jets) *SR&T* (flexible body effects) | *ATS F* (high-power arrays) (slip rings) | *ATS F* and *G* (30 ft deployable array) | *SR&T* *ATS V* (200 W TWT 12 GHz) (multipacting high-power PF components) | *ATS F* | *ATS V* *ATS III* (SHF and millimeter-wave propagation) | *SR&T* *ATS F* (small and transportable) | *ATS V* | *ATS F* *SR&T* (small, low cost) | *SR&T* *ATS III* *ATS F* ($100 converters) |
| Mobile air traffic control (surveillance, communication, and positioning) | Aircraft VHF *SYNCOM III* *ATS III* (ARINC) *ATS F* (place) | *ATS III* (address codes) *ATS IV* *SR&T* (coding and modulation) (error-correcting codes) | *ATS F* (reaction wheel) (dual spin) | *ATS F* (two-axis array drive) | *ATS III* *ATS F* | *SYNCOM III* *ATS I* *ATS III* (onboard processing) (40 W L-band) | *SYNCOM* *ATS I* *ATS III* *ATS F* | *SYNCOM* (VHF) *ATS III* (UHF) *ATS F* (L-band) | *SYNCOM III* *ATS I* (ARINC) *ATS III* (OPLE) | *ATS III* *ATS F* (modems) | *ATS F* *ATS I* *SR&T* (low-drag designs) | *ATS F* *SR&T* (low-noise detectors) |

*Source*: Midwest Research Institute 1971, 47–48.

## NASA's Contribution: Cost-benefit Analysis

The history of satellite communications technology (s.c.t.) and of NASA's role suggests that it may be useful to distinguish between two periods: the period prior to the first commercialization of s.c.t. in 1965 and the period after 1965.[5] We will argue that NASA's contribution was *decisive* in the first period and therefore that the direct benefits from the s.c.t. of 1965—henceforth termed the baseline satellite communications technology, b.s.c.t.—should be *fully* attributed to NASA. A preliminary calculation of these direct benefits, for the first submarket—telephony between the United States and Europe—will follow. After 1965, the direct contribution of NASA seems to be rather small, relative to its contribution in the previous period and probably to the contribution of the companies directly participating in the commercial exploitation of s.c.t. NASA's contribution is basically an indirect one—of having been responsible for the b.s.c.t., upon which the dramatic improvements in s.c.t. occurred, a process which enabled the gradual diffusion of this technology among an increasingly wide spectrum of applications. It will be argued that existing cost-benefit analysis is, by itself, inadequate to evaluate these benefits, the reason being our lack of an adequate conceptual framework which will consider (1) the timing of introduction of new technologies by the private sector, (2) the competitive reaction of existing technologies to the threats posed by new technologies, and (3) the external economies which any application provides to those firms engaged in subsequent applications.

### NASA's role in the initial applications

Without NASA's involvement, the introduction of the b.s.c.t. would have been delayed, at least beyond the date where its efficiency was surpassed by that of the alternative cable technology.[6] The technological history of NASA's involvement in s.c.t. is proof of the criticality of this involvement for the emergence of the b.s.c.t. The opinion of experts prior to NASA's involvement and after failure of the *ADVENT* program was overwhelmingly pessimistic about the prospects of a commercially viable synchronous satellite communications system. The MRI study (carried out in 1970) states: "Only slightly more than 10 years ago, many experts had serious doubts that: satellites could survive in space, and operate long enough to pay out; the quality of satellite communications transmissions would be acceptable; and that the cost of satellite systems would be competitive with traditional earth-based communication" (1971, 8).

---

5. This corresponds to Enos's (1962) distinction between the $\alpha$-phase and the $\beta$-phase of the innovation process of an industry. The $\alpha$-phase corresponds also to Kuznet's (1975) initial application phase in the innovation cycle and to NSF's (1973) "innovative" phase.

6. That is, beyond 1970, the year when the TAT-5 cable was laid across the Atlantic.

This outlook was largely responsible for Dr. Rosen's inability to "sell" the idea of a synchronous satellite to either a military or a civilian agent; NASA's taking over of the project was essential for the successful development of s.c.t., at least during the 1960s.[7]

Some quantitative evidence on expenditures on R & D and engineering efforts also bears upon the previous conclusion. NASA's expenditures on s.c.t. between 1960 and 1965 dwarf that of any other agency or firm potentially interested in a viable commercial system. Hughes's accumulated expenditures prior to NASA's taking over the synchronous satellite project amounted to approximately $1 million.[8] The accumulated expenditure of NASA in the *SYNCOM* project was probably of the order of $68 million. There was no chance whatsoever that Hughes would have committed itself to develop the technology without government support.[9] Overall, the expenditures of NASA on the *RELAY, ECHO*, and *SYNCOM* programs for the period including 1965 amounted to over $128 million.[10] The only serious contender with NASA in terms of private effort expended on s.c.t. is probably AT&T, in relation to *TELSTAR*. NASA cooperated with AT&T in this program, but the amount of government support is assumed to be low.[11] The cooperation took the form of using the NASA-constructed ground station—built for *RELAY*—in the *TELSTAR* program, and was mutual. While no other dollar figure for the cost of *TELSTAR* is available, it is unlikely to have represented more than a quarter or one-third of the NASA expenditures in s.c.t., until (and including) 1965. In any case, AT&T had serious doubts at the time about the viability of a synchronous s.c. system,[12] and it is highly unlikely for this and other reasons that it would have undertaken the effort required to arrive at a commercially viable system in case NASA would have failed to "adopt" Rosen's idea. In short, there seems to be little doubt that NASA's involvement accelerated the introduction of a commercial satellite communications system by at least five years.

7. Interview with Dr. H. Rosen, July 1978.
8. Interview with Dr. H. Rosen, July 1978.
9. Interview with Dr. H. Rosen, July 1978.
10. The data were obtained from the project development plans of Goddard Space Flight Center. Total projected expenditures on the three programs (including those beyond 1965) amounted to $130.7 million, distributed as follows: *ECHO* $19 million; *RELAY*, $43.6 million; and *SYNCOM*, $68 million. Costs include the costs of the spacecraft, launch vehicle, ground operations and support, and experiments. The costs of *SYNCOM* include $29 million on ground stations presumably paid by DOD and not by NASA. MRI reports that total NASA spending on communications satellites prior to the *ATS F* and *G* satellites amounted to $137 million.
11. Interview with Alton Jones of Goddard Space Flight Center, October 1978.
12. Especially about the effects of delay and echo on the quality of voice communications (see first section).

Direct benefits from NASA involvement: A calculation

The immediate benefits of NASA's involvement are the resource savings realized in the first application of the b.s.c.t, the United States–Europe telephony submarket. These resource savings are calculated for the period in which the b.s.c.t. was more efficient than installed cable technology, and discounted back to 1965. The amount of savings in any year is assumed to be the savings per circuit *installed* multiplied by the amount of satellite circuits *in use* during that year. The actual calculation is based on the following data, criteria, and assumptions:

1. Costs: *INTELSAT I* (the b.s.c.t.) investment costs per circuit of capacity compared with the comparable costs of the TAT-4 cable. This excludes a series of additional costs such as operation and maintenance costs, administration costs, ground station investment costs (for the satellite only), tails (for cable only), and possibly profits and taxes. Lasher and Owen's figures for the share of satellite investment costs over total costs for *INTELSAT* were 42% and 43% for the TAT-5 and TAT-6, respectively.[13] Therefore, use of satellite investment costs and cable investment costs in our calculation is probably indicative of the relationship between total satellite costs and total cable costs.[14]

2. Period: The relevant period ends with the earlier of two dates—the expected date of introduction of the b.s.c.t., under the assumption of no NASA involvement, and the actual date when cable technology in use became more efficient than the b.s.c.t. Table 10.3 shows the relative efficiencies of the various TAT cables. The design life of these cables is 20 years. The efficiency of the *INTELSAT I* technology is given by:[15]

> Investment, $9,000,000
> Design life, 1.5 years
> Circuit capacity, 240
> Investment cost per capacity circuit, per year, $25,000

To find the present value (PV) of the investment cost per satellite capacity circuit for a 20-year period—the cost figure which has to be compared with unit circuit cable costs of the last column of table 10.3—we take the PV of $25,000 per year for a period of 20 years discounted at 10% interest rate,

$$PV = 25,000 \; \frac{(1 - (1/1.1)^{20}}{0.1} = 213,000.$$

13. On a present-value basis. See Office of Telecommunications Policy 1971.

14. We exclude, however, redundancy in our calculation of satellite investment costs although in a sense we take the actual frequency of launch failure into account (it was zero).

15. See Midwest Research Institute (1971), 56. An alternative figure is $30,000 (Edelson, Wood, and Reber 1975).

Table 10.3. Relative efficiencies of undersea cables

| Cable | Year | Length (nautical miles) | Initial investment | Circuit capacity | Investment cost per alternate circuit |
|-------|------|-------------------------|--------------------|--------------------| -------------------------------------|
| TAT-1 (SB) | 1956 | 2,268 | 44.9 | 48 | 880 |
| TAT-2 (SB) | 1959 | 2,531 | 42.0 | 48 | 875 |
| TAT-3 (SD) | 1963 | 3,518 | 46.4 | 141 | 329 |
| TAT-4 (SD) | 1965 | 3,599 | 46.0 | 138 | 333 |
| TAT-5 (SF) | 1970 | 3,441 | 87.0 | 845 | 103 |
| TAT-6 (SG) | 1976 | 3,692 | 191.0 | 4,000 | 48 |
| TAT-X (SH) | 1985[a] | 3,600[a] | 290.0[a] | 12,000[a] | 24[a] |

Source: From Ramji 1976.
[a] Estimated.

A comparison of the above figure with the last column of table 10.3 shows clearly that (1) b.s.c.t. is *more* efficient than cable technology at the time of its commercial application (1965); (2) it is *less* efficient than the subsequent generation of cable technology, which came into use in 1970. Since our previous discussion concluded that the b.s.c.t., with no NASA involvement, would have been introduced after 1970, we take the latter date as our end year for the calculation of the impact effect of NASA's involvement. The period for our calculation is therefore 1965–1970.

3. Volume of traffic: Table 10.4 gives figures for satellite circuits in our submarket during 1965–1970. We decided to calculate resource savings over the volume of satellite traffic *in use* rather than over satellite circuit *capacity*. The former would be closer to the amount that TAT-4 cables would have had to substitute had the b.s.c.t. not been available. The calculation based on satellite circuit capacity is also made for reference purposes. Some additional assumptions are implied when using this measure: (1) Satellite circuits in use during the period would not have been substantially fewer had no improve-

Table 10.4. Satellite circuits: United States – Europe telephony submarket

| | 1965 | 1966 | 1967 | 1968 | 1969 | 1970 |
|-|------|------|------|------|------|------|
| Circuits in use[a] | 64 | 72 | 165 | 242 | 464 | 560 |
| Additional circuits in use | 64 | 8 | 87 | 77 | 222 | 96 |
| Capacity[b] | 240 | 240 | 240 | 1,500 | 1,500 | 6,000 |
| Additonal capacity circuits | 240 | 0 | 0 | 1,260 | 0 | 3,470 |

[a] From Ramji 1976.
[b] From Owen 1976.

Table 10.5. Resources saved by satellite communications technology in its first application, 1965–1970

| Year | Additional satellite circuits in use[a] | Additional capacity circuits[a] | Resources saved[b] | | Discount factor[c] | Present value of resources saved[d] | |
|---|---|---|---|---|---|---|---|
| | | | (1) | (2) | | (1) | (2) |
| 1965 | 64 | 240 | 7.68 | 28.8 | 1 | 7.68 | 28.8 |
| 1966 | 8 | 0 | 0.96 | 0 | 1.1 | 0.87 | 0 |
| 1967 | 87 | 0 | 10.44 | 0 | 1.21 | 8.62 | 0 |
| 1968 | 77 | 1,260 | 9.24 | 151.2 | 1.33 | 6.94 | 113.68 |
| 1969 | 222 | 0 | 26.64 | 0 | 1.46 | 18.24 | 0 |
| 1970 | 96 | 3,740 | 11.52 | 448.8 | 1.61 | 7.15 | 278.75 |
| Total for period | | | | | | 49.50 | 421.23 |

[a]See table 10.4.
[b]In millions of dollars. The computation is as follows: $120,000 times additional satellite circuits in use for (1); $120,000 times additional satellite capacity circuits for (2).
[c]At $i = 10\%$, it is $(1.1)^{t-1965}$.
[d]Resources saved divided by the discount factors.

ments beyond the b.s.c.t. occurred; (2) the additional cable circuits of the TAT-4 type that would have been installed in the absence of the b.s.c.t. are roughly equal to the satellite circuits actually used during the period.[16]

4. Unit resource savings: Point 2 above implies[17] that our calculations implicitly are made on the assumption that cables are perfectly divisible in capacity while indivisible in lifetime. This is unrealistic but enables us to avoid the problem of having to decide the timing of introduction of the additional TAT-4 technology cables which would have been laid had the b.s.c.t. not become available. In any case, it is doubtful whether our calculations would have been significantly altered had that decision-making been explicitly taken into account. Under our assumptions, 20-year circuit resource savings, discounted, equals the difference between (a) unit circuit costs with TAT-4 technology, which would be paid at the time the additional cable would have been installed, and (b) PV = 213,000. This amounts to $120,000.

*The Calculation:* The above elements of analysis enable us to compute the direct impact of NASA's involvement in satellite communications technology in its first commercial application: telephony along the United States–Europe route. Table 10.5 indicates the procedure followed. Note that the resources saved in any particular year depend on the additional satellite circuits used or

16. If (1) does not hold, then part of the direct benefits which we ascribe to the b.s.c.t. would really be the result of post-b.s.c.t. technology.

17. Over and beyond the effects of the price elasticity of demand for telephone circuits in the main Atlantic route.

installed during that year and not on the absolute number of circuits.[18] The present value of the resources saved is $49.5 million when the calculation is based on satellite circuits in use during the period, and $421.23 million when it is based on capacity. $49.5 million is, in our opinion, a lower bound to the resources saved from the first application of the b.s.c.t. There are a number of factors which would raise the figure:

1. We have assumed that cable technology was independent of satellite technology. Experience tells us that developments in cable technology were considerably accelerated by the introduction of satellite technology. This would tend to lengthen the period when the benefits from the b.s.c.t. should be computed.

2. The unit circuit savings from the availability of the b.s.c.t. are higher than computed because of lower design life of satellites relative to cables. Thus, after 1.5 years of life, a b.s.c.t. circuit could be substituted for a more efficient satellite technology, and so on throughout the 20 years of cable life. Part of these additional savings should be attributed to the b.s.c.t. because they would have occurred even without additional R & D and engineering resources explicitly allocated to improve the b.s.c.t., i.e., through operating experience, learning-by-doing in the production of hardware, and unpriced externalities received from other sectors.[19]

In our opinion, both of these factors certainly outweigh by far any possible overestimate in the volume of satellite circuits in use which would have been substituted by cables in the absence of the b.s.c.t.

### Conclusions

The direct benefits from introducing the b.s.c.t. into its first application—the United States-Europe telephony submarket—represent an important share of all the NASA expenditures leading to this technology. This share is at least a third, and probably much more. This is a remarkable achievement for the following reasons: (1) Satellite communications technology is a major technology or innovation[20] of post-World War II industrial society. It is sufficient to look at the set of actual and potential applications of this technology in order to realize this fact.[21] (2) The full impact of this (like any) major tech-

18. This is because we calculate savings per 20-year circuit rather than savings per 1-year circuit.

19. The theoretical issue here is the allocation of the resource savings from substituting an older technology with a long design life for a new technology with a relatively short design life between two stages in the history of the new technology: the baseline technology and the post-baseline improvements.

20. For the definition of a major innovation, see Kuznets 1971, 314–33.

21. See table 10.6, which is taken from the MRI (1971) study. Additional applications, potential or at various stages of realization, have since become relevant.

Table 10.6. Demand categories for major systems

| International satellite system (point to point) | Broadcast distribution (point to multipoint) | Direct broadcast satellites (point to wide areas) | Air traffic control (surveillance, communications, and positioning) |
|---|---|---|---|
| International telephone trunking | Network TV to local rebroadcast stations | Remote village TV; direct to home TV | Aircraft communications |
| Transoceanic television | School district ETC | Schoolhouse TV | Traffic control |
| Global multiple-access switching network | Rebroadcast ETV | | Position location |
| Commercial teletype, facsimile, data transmission | Larger CATV systems | | Geodesy |
| | Cable TV systems | | Data collection and relay |
| | Regional telecommunications | | Satellite-to-satellite relay |
| | High-quality high-speed facsimile | | |
| | One-way digital data distribution; financial information | | |
| | Timesharing services | | |
| | Meeting and polling | | |
| | Information networks | | |
| | Law and justice | | |
| | Medical and scientific | | |

nology on economy (and society) will make itself felt only after a reasonably long period of time.[22] The direct impact analyzed previously refers only to one submarket corresponding to one application, i.e., part of international telephone trunking, the first application shown in table 10.6. We doubt that we can find cases where the first direct impact of the baseline technological level of a major innovation covered one-third or one-half of the costs within six years.

NASA's overall contribution may be divided into three parts: (1) the resource savings due to the b.s.c.t. in the first submarket and in subsequent submarkets or applications penetrated by it; (2) the share of total resource savings due to post-b.s.c.t. which should be attributable to NASA; (3) the share of the value to society derived from the satisfaction of previously unmet

22. The full impact of Watt's steam engine, for example, took more than 100 years to manifest itself. Even allowing for shorter lead times in the various applications of current major innovations, a generation or two may still be required for the full impact to be felt in the economy.

needs. The first part is calculable and quantitative; it probably does not go very much beyond the international telephony submarket, because improvements beyond the baseline technology were essential for most of the subsequent diffusions of s.c.t. Cost-benefit analysis can be appropriately used here. The second and third components are extremely difficult to calculate for the following reasons. The technology beyond the b.s.c.t. is due to the joint action of the following factors or agents: NASA as the agent responsible for the introduction of the b.s.c.t., upon which subsequent improvements are based,[23] NASA as a contributory agent to post-b.s.c.t.,[24] commercial firms, INTELSAT and COMSAT, as contributors to post-b.s.c.t. The diffusion of post-b.s.c.t. within and across submarkets and applications depends on technology adoption and switching decisions of private firms. These decisions are influenced by the pattern of unpriced benefits (externalities) generated in the various applications, especially the initial ones. Moreover, the third component requires some radically new tools of analysis, presumably a theory of needs.[25]

Our empirical and theoretical knowledge of these processes is slight. This knowledge is essential for evaluating the impact of *major* innovations and the possible share of this impact which may be attributable to the government expenditure responsible for first commercialization. It is required *prior* to the application of cost-benefit tools of analysis. Alternately, its lack is the main reason why cost-benefit analysis cannot perform the job.[26]

### Issues Emerging from NASA's Involvement in Communications Satellite Technology

NASA's strategy and success in inducing commercialization of communications satellite technology

Success in the transition from early R & D (and baseline technology development) to commercialization can be ascribed to a number of factors. First, a

23. For this NASA should at least be attributed a "fair" rate of return to its investment in the b.s.c.t.

24. For example, as a result of the *ATS* program

25. If satellite technology meets a previously unsatisfied need (e.g., providing telephone services to a region for the first time), then by definition it does not save on resources previously devoted directly to that need. Benefits should be estimated as the value of meeting such a need from society's point of view. This may be a very difficult job.

26. It is not surprising, therefore, that we have found no real cost-benefit analysis of satellite communications technology as a major innovation. The MRI study, for example, refers to measures of direct economic impact. These include investments and revenues of COMSAT, growth of the system (ground stations and channel capacity), size of the market for hardware, etc. Case studies of how technological capabilities of firms augmented by their interaction with NASA work were subsequently applied to commercial communications markets are also recorded. However, the basic methodological issues for a cost-benefit analysis have apparently not yet been addressed.

substantial portion of the R & D and prototype development and building was performed by subcontractors such as Hughes which subsequently supplied the first (and subsequent) user(s).[27] To a significant extent there was no need for explicitly transferring the design and manufacturing capabilities from government to commercial use. Second, experimentation by NASA with a variety of different satellite systems prior to commercialization led to a cost-effective first commercialization of this major new technology. In this respect it is highly significant that NASA attempted c.s. programs of its own, over and beyond the support it provided to the programs undertaken by industry. The additional R & D widened considerably the range of technologies available for commercialization and considerably reduced costs. Without NASA's support, first commercialization would probably have embodied AT&T's[28] low-altitude system involving large numbers of satellites instead of the enormously cheaper synchronous satellite alternative involving only up to three satellites. Third, the fact that NASA sponsored *applications R & D* after having arrived at the basic design solution to s.c. (synchronous satellites) is important for two reasons: first, there are indivisibilities in this research, since it is more economical to undertake a very wide spectrum of applications experiments together in the context of a given program than it is to perform a series of individual experiments, each one suited to the needs of a particular customer; second, there are spin-offs (externalities) flowing from research primarily useful for some applications that benefit other applications. These will be taken into account by a government agency like NASA but need not be taken into account by private firms. An implication of these two characteristics of applications R & D is that a NASA-sponsored program could be much more effective than a set of equivalent private-sector-sponsored programs.[29]

Another aspect of NASA's activity concerns the accumulation of a knowledge base for engineering design. This is an effective way of stimulating the commercialization of a technology since it may considerably reduce the cost of finding an optimum point design for a particular application. Thus, the accumulated data base represents a knowledge infrastructure element to the

27. This was the case not only with respect to the basic design solution arrived at—*SYNCOM*—but also with respect to the subsequent applications R & D (*ATS*). The capabilities developed in the latter were also located in the contractor firms and had immediate commercial implications. It is worthwhile to note that Hughes's work on INTELSAT satellites led them to offer a "standard" satellite, the type HS-33, which was the basis for the *WESTAR* and *PALAPA* systems.

28. The only private sector firm capable of commercializing a c.s. system on its own was AT&T, which was working on a nonsynchronous system based on *TELSTAR*-type satellites. The company apparently believed that a synchronous system would not be economical for some years. See Smith 1976.

29. Less costly and with a stronger commercial impact (the latter a result of a higher rate of diffusion of c.s. technology).

industry, which should be differentiated from tangible capital infrastructure components. In s.c. a substantial accumulation of knowledge accompanied the *RELAY* and *SYNCOM* programs. They enhanced understanding of the (high-energy particle) space environment and enabled the optimization of communication and other subsystems.

### Division of labor with the private sector

Prior to proof of feasibility of a basic design solution, expected private benefits may be low and risks high, so expected private utility of R & D may easily be negative. Expected social utility, however, may be positive owing to externalities and lower social risks relative to private risks. After a basic design solution is arrived at, the private profitability (commercial risk) of subsequent R & D whose objective is to optimize, adapt, and improve such a design increases (decreases).[30] Under these conditions, "baseline technology" R & D should be sponsored by the government, while subsequent R & D should be left to the private sector.

The above reasoning would have led NASA to launch *RELAY* and, once its feasibility was proved, to improve upon it in conjunction with the private sector. But why did NASA undertake the *SYNCOM* program, which led to a more effective basic design solution? The answer lies in the gap created between the social and private expected utility once the synchronous satellite idea became available. Given that there already existed a proved basic design (*RELAY*-type active satellite systems) and that the new proposed design was very risky for private firms, the latter would not have undertaken the project, at least as early as NASA did. However, the expected social utility of the proposed new basic design was probably very high, owing to the relatively lower social risks and to externalities. Thus, it may have made sense, ex ante, for NASA to attempt the development of a more advanced technology even before the full potentiality of the existing nonsynchronous s.c. was fully realized.

Federal support for s.c. technology may be justified even beyond the stage of first commercialization, for a variety of reasons. First, private firms may have a tendency to perform low-risk, incremental R & D only and may neglect longer-term research on technologies, even when these offer a positive social utility. In these cases, governments should step in. The current conservatism of the satellite communications commercial sector and potential for significant

---

30. Proof of feasibility of a particular basic design solution (e.g., of synchronous c.s.) enables a sharper specification of subsequent technological needs in a variety of areas. Needs can now be expressed in terms of required improvements in particular subsystems of the corresponding product class rather than in general terms. This has been termed "stronger need determinateness" or "focusing devices" (the latter when performance limitations of a particular subsystem are the effective bottleneck to overall system improvement). See Teubal (1979) and Rosenberg (1976).

(if risky) advance are observed by P. L. Bargellini of COMSAT in presenting a table of potential improvements based on new technologies. He comments:

> Table 3, which summarizes the situation, reveals at least four areas in which the new technology has not been tried. Since some of these technologies appear quite promising, it is appropriate to hope that the impasse between the requirements of operational systems and the need for experiment will be broken. The sooner this impasse is overcome the better; however, the success of "conventional" satellite systems and spacecraft design appears to have been partially responsible for slowing down experimental programs. This situation is not surprising, since it has occurred many times in the past in other fields of technology where large investments in operational systems retarded the introduction of subsystems and components based on new technologies, notwithstanding the potential promise.[31]

Second, the gap between private and social expected utility (including the risk and benefit appropriability factors) of developing new technologies may conceivably increase during depressions. The cash-flow problems of firms may deter them more than in normal times from undertaking R & D. On the other hand, the rapid introduction and diffusion of new technologies may be critical to overcome the depression. On both counts the role of government-sponsored R & D may have to be enhanced during depressions rather than being reduced.

Finally, foreign governments' continued support of certain technologies may justify, in certain cases, enhanced (and possibly redirected) public support at home. We do not know of any rigorous criteria for identifying these cases in practice.

### Relations with users and effects on market structure

NASA's support of s.c. R & D had a major effect on the structure of the industry, and this is because of the variety of roles it played within the innovation process. If NASA had simply subsidized part of the ongoing R & D efforts of industry, then it is likely that the domination of AT&T in both domestic and international communications would have continued. This is because AT&T was the only private company which had already undertaken a considerable R & D effort of its own and was capable of proceeding through commercialization of its nonsynchronous system. The fact that NASA—beyond supporting *TELSTAR*—undertook programs of its own such as *RELAY* and more importantly the successful *SYNCOM* program (based on the Hughes idea of synchronous satellites) opened up both the communications and the satellite and communications equipment markets. Thus, transatlantic telephone

---

31. Bargellini 1978, 39. The "competition" between conventional and new technologies is one of the main themes explored in Teubal and Steinmueller 1984.

via satellite turned out to be owned and operated by an international consortium—INTELSAT—and the United States part by a private/public corporation with a relatively broad-based ownership (COMSAT). Similarly, the technological preconditions were created for the FCC's 1972 decision assuring multiple-entry into the United States domestic s.c. market. (Finally, NASA contracting created hardware and system expertise in a wide variety of firms such as Hughes, RCA, etc.)

The first commercial application or "use" of s.c. was the transmission of telephone and TV signals across the Atlantic. In the early 1960s this was the only possible commercial application given the high cost of the existing medium- and low-altitude satellite systems, e.g., based on AT&T's *TELSTAR*. The existing supplier of telephone services in this market was AT&T (via underwater cables) and the eventual supplier was INTELSAT. AT&T held a 29% share of COMSAT, the company representing the United States in that consortium. In addition to the suppliers of s.c. services ("users of the technology"), we have the "final users" on both sides of the Atlantic represented by a variety of government agencies. From all this it is clear that the issue of NASA's relationship with the users is a complex one. Some user-related aspects which were important for the success in the transfer or commercialization of the technology include (1) support of the R & D programs of existing suppliers of services (i.e., of AT&T's *TELSTAR* program) without neglecting experimentation with other technologies; (2) increasing cooperation with European countries to coordinate experimentation. This started with *ECHO* and continued with *RELAY*. In 1961 the telecommunication agencies of the United Kingdom and France agreed to build the ground stations for these experiments. This pattern of NASA-sponsored international cooperation with the final users of the first commercial system enormously accelerated and facilitated the establishment of INTELSAT. Moreover, the chain of ground stations built was one of the foundations of the global system (see Smith 1976) and presumably explains the speed of the technology transfer process.

### Concluding remarks

The paper is an attempt to evaluate the contribution of NASA to the emergence of the s.c. industry. It combines aspects of the qualitative history of the technology and a partial quantitative analysis of the costs and benefits of NASA's involvement. Prior studies, to our knowledge, have generally emphasized one or the other, so it may be useful to consider the advantages of combining both. Specific features of the work include:

1. A distinction is made between base-line technology development and subsequent technological development. The distinction is important because the qualitative case histories show that the contribution of NASA is clear and direct in the former and indirect in the latter. Therefore, a partial and relatively

standard cost-benefit analysis of NASA's involvement appears as a possibility. Our present tools of analysis do not permit us, however, to evaluate NASA's share of the social benefits accruing after 1970.

2. The cost-benefit analysis is an ex post analysis, and it attempts to measure *actual* costs and benefits but cannot answer the issue whether NASA decision making, ex ante, was justified in the light of the information available when the decisions were made. In this sense it is similar to Henderson's (1977) study of the Concorde and the United Kingdom's advanced gas-cooled reactor. This is the main reason for not considering a risk premium in the discount factor applied to the benefits.[32]

3. The cost-benefit analysis assumes that NASA's involvement *accelerated* the introduction of the new technology. This has been recognized as the major effect of government involvement in high-technology projects (Henderson 1977; Pavitt and Walker 1976), but we have not seen cost-benefit analysis that takes this into account explicitly. (This requires, of course, a good knowledge of the qualitative history of the technology.)

The s.c. case is an example of successful government intervention in what has become a successful new industry. It may be interesting to compare it with other less successful cases, such as Concorde or the advanced gas-cooled reactor. We will restrict ourselves, however, to comparing some of its salient features with those found in Keck's (1980) study of the West German nuclear reactor industry. Government policy was apparently less successful there than in satellite communications, at least from the viewpoint of its direct effects. For example, in contrast to s.c., the major technologies supported by the government were not commercialized, and government support of those that were was relatively minor. Correspondingly, the effect of government support in accelerating the introduction of the new technology was much less significant than in the s.c. case.[33] Similar differences are apparent with respect to the transfer of the technology to the sector involved in commercialization. In the case of s.c., the major contractors (e.g., Hughes) collaborating in the design and building of hardware for NASA actually supplied the first commercial user of the technology (INTELSAT). Thus transfer was almost automatic.

The situation is far less clear in the West German nuclear reactor case, since many government contractors left the business before constructing commercial units. Moreover, there really was no significant technology transfer issue since firms like Siemens which first commercialized did not base their technology on

---

32. There are also theoretical reasons for *not* including a risk factor in ex ante evaluations of *public* expenditure projects. See Arrow and Lindt 1970, 364–78.

33. The indirect effects of government support may have been significant but these may be extremely difficult to trace; for example, manpower support, the effect of weeding out noneconomical technologies.

contracts with the government but on their own R & D efforts. Finally, the user of the new technology in the West German case was a relatively clearly defined entity—utilities or combinations of utilities-and its role in choosing the technology to be used was paramount. In neither case is this the situation with respect to s.c. Probably because of very special institutional and technological characteristics of the industry, NASA and other United States agencies played a much more active role both in arriving at the optimal baseline technology and in determining the particular nature of the technology user who would own and operate the first system on a commercial scale. These characteristics may also limit the applicability of NASA's experience to other instances of government support of large high-technology projects.

## References

Arrow, A., and R. Lindt. 1970. "Uncertainty and the Evaluation of Public Investment." *American Economic Review*, pp. 364-78.

Arrow, K. 1961. "Economic Welfare and the Allocation of Resources for Invention." In *The Rate and Direction of Inventive Activity*, edited by R. Nelson, 609-25. Princeton: Princeton University Press.

Bargellini, P. L. 1978. "A Review of U.S. Satellite Communications Technology." COMSAT.

Burns, A. 1934. *Production Trends in the U.S. since* [1910]. New York: National Bureau of Economic Research.

Edelson, B., H. Wood, and C. Reber. 1975. "Cost Effectiveness in Global Satellite Communications." Paper presented at the I.A.F. 26th Congress, September.

Enos, J. 1962. *Petroleum, Progress and Profits*. Cambridge: MIT Press.

Hartman, E. P. 1970. *Adventures in Research*. NASA

Henderson, P. H. 1977. "Two British Errors: Their Probable Size and Some Possible Lessons." *Oxford Economic Papers* (July 1): 159-205.

Keck, Otto. 1980. "Government Policy and Technical Choice in the West German Reactor Programme." *Research Policy* 9:302-56.

Kuznets, S. 1930. *Secular Movements in Production and Prices*. Clifton, N.J.: Kelley.

Kuznets, S. 1971. *Economic Growth of Nations*. Cambridge: Harvard University Press.

Kuznets, S. 1975. "Technological Innovation and Economic Growth." In *Technological Innovation: A Critical Review of Current Knowledge*, edited by P. Kelly and M. Kranzberg. San Francisco: San Francisco Press.

Midwest Research Institute. 1971. "Economic Impact of Stimulated Technological Activity, Part II—Case Study—Technological Progress and Commercialization of Communications Satellites." MRI no. 3430-D. NASA Contract NASW-2030. October 15.

Mueller, D. C., and J. E. Tilton. 1969. "Research and Devlopment Costs as a Barrier to Entry." *The Canadian Journal of Economics/ Revue canadienne d'économique* 11 (4): 570-79.

National Science Foundation. 1973. *Interaction between Science and Technology in the Innovation Process.* Washington D.C.

Office of Telecommunications Policy. 1971. "International Facilities Study."

Owen, B. 1976. "The International Communications Industry, the TAT-6 Decision, and the Problem of Regulated Oligopoly." Department of Economics, Stanford University.

Pavitt, K., and W. Walker. 1976. "Government Policies towards Industrial Innovation: A Review." *Research Policy* 5:11–97.

Pickard, R. H. 1963. *Relay I Spacecraft Performance.* Publications of Goddard Space Flight Center. II. Space Technology. Greenbelt, Md.: GSFC.

Pickard, R. H. 1965. *Final Report on the Relay I Program.* Greenbelt, Md.: GSFC.

Pritchard, W. L. 1977. "Satellite Communication—An Overview of the Problems and Programs." *Proceedings of the IEEE* 65 (3): 294–308.

Ramji, S. 1976. "The Role of Satellites and Cables in Intercontinental Communications." Paper presented at the Sixth AIAA Communications Satellite Systems Conference, Montreal, Canada.

Redisch, W. M. 1975. "ATS-6 Description and Performance." *IEEE Transactions on Aerospace and Electronic Systems* 11 (6).

Rosen, H. 1963. "Synchronous Communication Satellite." In *Space Communications*, edited by A. V. Balakrishnan. New York: McGraw-Hill.

Rosenberg, N. 1976. "The Direction of Technological Change: Inducement Mechanisms and Focusing Devices." In *Perspectives on Technology*, edited by N. Rosenberg, 108–25. Cambridge: Cambridge University Press.

Smith, Delbert. 1976. *Communication via Satellite: A Vision in Retrospect.* The Hague: A. W. Sitjoff.

Teubal, M. 1979. "On User Needs and Need Determination: Aspects of the Theory of Technological Innovation." In *Industrial Innovation: Technology, Policy, Diffusion*, edited by M. Baker, 266–93. London: Macmillan. (Chap. 4 in this volume.)

Teubal, M., and E. Steinmueller. 1984. "The Introduction of a Major New Technology: Externalities and Government Policy." In *The Economics of Relative Prices*, edited by B. Czikos-Nagy, D. Hague, and G. Hall, 117–39. London: Macmillan. (Chap. 11 in this volume.)

Funds for the support of this study have been allocated by the NASA-Ames Research Center, Moffett Field, California, under Interchange no. NCA2-OR745-815. We are especially grateful to Harold Hornby for his advice and help in defining a project on the economic impact of federal R & D and to Harry Goett for having suggested taking satellite communications as a case study. Harry Goett has also given generously of his time and provided important advice and encouragement. For all this we are extremely grateful. Our thanks to P. Bargellini, L. Jaffe, A. Jones, and Professor Lusignan for their cooperation at various stages of our work and to J. White, H. Holley, G. Allen Smith, D. Smith, J. Hart, and Dr. R. T. Jones, who provided useful background information on the activities and impacts of the National Aeronautics and Space Administration. We appreciated the comments of two anonymous referees, who suggested improvements in an earlier draft.

# 11 The Introduction of a Major New Technology
## Externalities and Government Policy

## Morris Teubal
## and Edward Steinmueller

The usual justification for government support of innovative activity is that the firms engaged in it generate externalities as a result of the special properties of information (imperfect appropriability, negligible costs of reproduction, etc.). These externalities lead an economy with competitive goods markets to under-invest in R & D (see Arrow 1962). Arrow's analysis does not distinguish between the generation of information, its embodiment as a new technology (invention), and its application to production (innovation). Such distinctions are implicitly or explicitly made by Barzel (1968) in his analysis of the firm's decision to introduce a new technology, and by policy-oriented research concerning the impact on the economy of the activities of a government agency such as NASA (see also Robbins, Kelley, and Elliott 1972; Mathematica, Inc., 1975). Barzel concentrates on the *timing* of introduction of an innovation or technology in cases where the basic knowledge used by the innovation is a public good. Robbins, Kelley, and Elliott (1972) and Mathematica, Inc. (1975), attempt to determine whether and how much NASA or other government agencies *accelerate* the introduction of innovations and sometimes to evaluate the associated economic benefits.

Furthermore, Barzel's paper does not consider the external effects of the first application of a new technology or innovation. His analysis and conclusions are therefore not applicable to the introduction of a *major* new technology.[1] A salient characteristic of such technologies is their relatively wide range of applications, both in the economic sector in which they are first applied and beyond it. The classical examples include the steam engine, the grinding machine,

1. For the present purpose, whether a technology is a major one should be assessed on the basis of the resources it saves. For an extensive analysis of the characteristics of major innovations see Kuznets (1971, 314–33), and Kuznets (1975).

transistors and other semiconductor devices, and computers.[2] Another characteristic of a major technology is that as it becomes diffused, each application generates experience and information which increases the profitability of subsequent applications.[3] Externalities are bound to be generated under these circumstances: it is unlikely that the innovator firm will be able to appropriate all the benefits, present and future, that will eventually derive from all applications of its innovation.[4] The appearance of key sectors or firms may be interpreted as an attempt to internalize at least part of these benefits.[5]

The object of this paper is to analyze the socially optimal introduction time of a major new technology that is potentially applicable to a number of economic sectors. Its main emphasis is to make explicit the distinction between the first and subsequent applications of the technology. It is assumed that in its first application the new technology not only reduces production costs but generates information which facilitates subsequent applications. The associated externalities cause a divergence between the private and the social optimum introduction times of the new technology.[6]

We begin by analyzing the private decision to introduce a major new technology. A discussion of the externalities generated follows, and we then proceed to determine the social optimum introduction date. We then go on to analyze the factors determining the size of the optimum technology-introduction subsidy and apply the framework to determine the social benefits derived from accelerating the availability of a major new technology.

## The Basic Model and Its Assumptions

We may distinguish three points of time in relation to the new technology: the data at which it becomes "available," the date of "introduction," and the date at which "diffusion" begins.

2. Watt's steam engine was initially applied in ironworks and later spread to the powering of industrial machinery, to land transportation (the railroad), and to water transportation. This process of intersectoral diffusion took over 50 years. For a list of applications of satellite communications technology see Midwest Research Institute 1971, 47–48.

3. This transpires from the histories of major technologies. Several examples are described by Rosenberg 1963.

4. A striking example is the transistor. Cumulated license fees were estimated at about £3 million from 1952 to 1963 (excluding cross-license benefits). See Freeman 1974, 142. This sum is certainly way below the social benefits derived from the innovation.

5. Key firms specialize in adapting a technology to a wide range of applications and selling them to firms in other areas. For a description of the emergence of machine tool firms in the United States in the nineteenth century, see Rosenberg 1963.

6. The divergence between the private and social optimum introduction dates analyzed by Barzel (1968), Kamien and Schwartz (1972), Scherer (1967), and other authors is caused by the existence of competition in introducing (or using) the innovation and not, as in this paper, by externalities. Our analysis does not consider the implications of competition among potential innovators.

*Availability.* A new technology is, in our context, considered available when its feasibility has been demonstrated (invention); and when it has become possible to estimate the costs and benefits of applying the invention with sufficient accuracy to warrant its inclusion in the set of technological options confronting entrepreneurs.

*Introduction.* This is the moment when the new technology is in fact selected from the set of options for the first time (innovation). It will occur when the introducing firm is prepared to incur the fixed cost of innovating, which implies that the new technology is expected to generate profits.

*Diffusion.* Diffusion begins when the technology spreads to a sector other than the one in which it was first applied. At this point, the information and experience gained by the innovating sector is available, at least in part.

Thus we consider an economy with just two sectors, sector 1 and sector 2. Given such an economy, the question arises through which of them the new technology will be introduced. This complex question is beyond our present scope; as far as this paper is concerned, the sequence of sectors using the technology is predetermined.

*Assumptions.* (A1) Operating costs when using the conventional technology are constant and stationary in both sectors. This implies that there is no firm-sponsored or any other R & D intended to improve the conventional technology. (A2) There is exogenous (e.g., government-sponsored) R & D activity which makes the new technology available at time $t_A$. (A3) From then on, this R & D continues to reduce the operating costs of the new technology but (A4) is assumed to have no further effect on the fixed cost of introduction, $M_I$, regardless of the time of introduction, $t_I$. Once this has occurred, (A5) the learning that takes place in the introducing sector 1 is assumed to have no effect on the sector's own operating costs with the new technology; but (A6) it will reduce the fixed cost of diffusion, $M_D$, until the time, $t_D$, at which the new technology spreads to sector 2. (Thereafter the only element affecting cost is the ongoing R & D which reduces operating costs in both sectors.)

We further assume that (A7) sector 1 consists of a single firm and that (A8) there is no threat from potential entrants to the sector. (A9) The sector's demand curve is stationary. (A10) The decision to introduce the new technology entails the abandonment of the conventional technology.

Assumptions (A1) and (A9) imply that sector 1's operating profits per unit of time when using the conventional technology are constant:[7]

$$\bar{\pi}_1(t) = \bar{\pi}_1. \tag{1}$$

---

7. Operating profits, $\pi$, is the quantity produced *times* the difference between price and unit operating costs; while not explicitly considering how quantity and price are determined in sector 1, it is likely that they will be stationary if both operating costs and demand are stationary.

Similarly, (A3) and (A9) imply that operating profits when using the new technology are increasing, i.e.,

$$\pi_1'(t) > 0. \tag{2}$$

The new technology is available at $t_A$ ($= 0$) and can be introduced at either of two times, the private and social optimum introduction times, respectively $t_p$ and $t_s$. In order to determine $t_p$, we should look at the conditions under which sector 1 will switch to the new technology (A10).

Let $\bar{\Pi}_1(t_I)$ be the present value at time 0 of operating profits when using the conventional technology from time 0 to time $t_I$. Similarly, $\Pi_1(t_I)$ will be the present value at time $= 0$ of operating profits from using the new technology, for all $\theta$ starting with $t_I$, net of the present value of the fixed cost of introduction, $M_I$. Then, with $\rho$ denoting the discount rate, we have

$$\bar{\Pi}_1(t_I) = \int_0^{t_I} \bar{\pi}_1 \exp(-\rho\theta)\,d\theta = (\bar{\pi}_1/\rho)\,[1 - \exp(-\rho t_I)], \tag{3}$$

where $\bar{\Pi}_1(0) = 0$, $\bar{\Pi}_1'(t_I) = \bar{\pi}_1 \exp(-\rho t_I) > 0$, and $\bar{\Pi}_1''(t_I) < 0$;

$$\Pi_1(t_I) = \int_{t_I}^{\infty} \pi_1(\theta) \exp(-\rho\theta)\,d\theta - M_I \exp(-\rho t_I), \tag{4}$$

where $\Pi_1(\infty) = 0$ and $\Pi_1'(t_I) = -\exp(-\rho t_I)\,[\pi_1(t_I) - \rho M_I]$.

It is clear that $\bar{\Pi}_1(t_I)$ increases with time, while $\Pi_1(t_I)$ increases or decreases according as the savings in the *flow* of fixed-cost services, $\rho M_I$, exceed or fall short of the operating profits, $\pi_1(t_I)$, lost. In terms of figure 11.1, this depends on whether $t < \hat{\imath}$ or $t > \hat{\imath}$.[8]

The optimum private introduction time, $t_p$, is that value of $t_I$ which maximizes

$$\bar{\Pi}_1(t_I) + \Pi_1(t_I) \equiv V_1(t_I). \tag{5}$$

The first-order condition[9] for max $V_1(t_I)$ is

$$V_1'(t_p) = \exp(-\rho t_p)\,\{\rho M_I - [\pi_1(t_p) - \bar{\pi}_1]\} = 0, \tag{6}$$

from equations (3)–(4). This implies that

$$\pi_1(t_p) - \bar{\pi}_1 = \rho M_I \tag{7}$$

and

---

8. Thus $\hat{\imath}$ is defined by $\Pi_1'(\hat{\imath}) = 0$, from which it follows that $\pi_1(\hat{\imath}) = \rho M_I$. Note that figure 11.1 assumes that $\Pi_1(\hat{\imath}) > 0$, i.e., that the ongoing exogenous R & D eventually leads to a profitable new technology.

9. Second-order conditions for a local maximum at $t_p$ always hold, as can be seen from $\bar{\Pi}_1''(t_p) + \Pi_1''(t_p) = -\exp(-\rho t_p)\Pi_1'(t_p) < 0$, and taking eq. (6) into account.

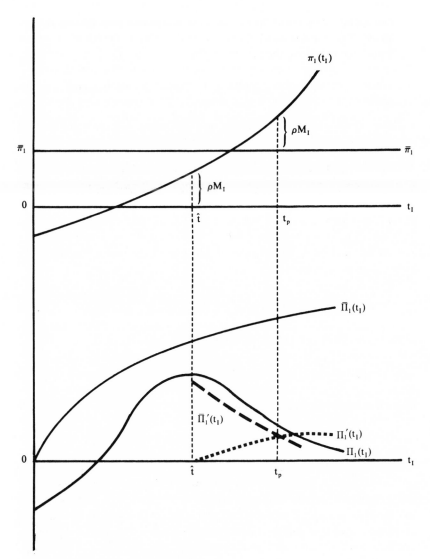

Fig. 11.1. Optimum private introduction time.

$$t_p > \hat{i} \tag{8}$$

(see fig. 11.1).

*Result 1*: At the optimum private introduction time, the difference in operating profits between the new and old technologies should be equal to the service flow of the fixed cost of introduction. Moreover, the greater the operating profits of the conventional technology (and presumably the more efficiently it is used), the farther in the future is the optimum private introduction time of the new technology.

*Result 2*: The optimum private introduction time will be later than that $t_I$ which maximizes the cumulated discounted profits of using the new technology.

These results explain why Alaska and Indonesia, for example, adopted domestic satellite communication systems before the United States.

### The External Effects of Introducing the New Technology

We start with a specification of our view that there is a predetermined sequence starting with sector 1: the experience from using the new technology in sector 1 is a prerequisite for diffusion to sector 2. We extend assumptions (A7)–(A10) to sector 2.[10] Let us denote the fixed costs associated with diffusion by $M_D$ and let $t_D$ be the time of diffusion of the new technology. The formulation we assume is

$$M_D = M_D(t_I, t_D), \qquad \frac{\partial M_D}{\partial t_I} > 0, \qquad \frac{\partial M_D}{\partial t_D} < 0. \qquad (9)$$

The specific formulation to be used later is

$$M_D = \frac{k}{t_D - t_I}, \qquad (10)$$

where $k$ is a positive constant. Note that

$$\lim_{t_D \to t_I} M_D = \infty$$

which is consistent with the assumption $t_D > t_I$.

The externalities generated by introducing the new technology are the additional profits and consumer surplus made possible in sector 2. The assumption that the price of sector 2 output is stationary implies that there is no additional consumer surplus, so that the externalities are equal to the change in profits at

10. As an alternative to the assumption of a single firm in sector 2, and following Barzel, we can distinguish between the owner and the users of the innovation. The owner would incur the fixed cost associated with use; he would charge royalties to users, thereby extracting all or part of the profits from diffusion. The date of diffusion would be chosen to maximize discounted profits net of fixed costs. Users may belong to a competitive industry provided diffusion is instantaneous (another of Barzel's assumptions). The analysis of our model would require only minor modifications when these conditions hold. Note that a similar set of conditions could replace our assumption of a single producer in sector 1.

the original output level (saving in costs).[11] Then letting $\bar{\pi}_2$ and $\pi_2(t)$ be the yearly profits of sector 2 under the old and the new technology, respectively, the externalities generated by introduction at $t_I$ and diffusion at $t_D$ can be expressed as follows:

$$E[t_D; t_I] = \int_{t_D}^{\infty} [\pi_2(\theta) - \bar{\pi}_2] \exp[-\rho(\theta - t_I)]$$
$$\times d\theta - M_D(t_D; t_I) \exp[-\rho(t_D - t_I)]. \quad (11)$$

The externalities are here expressed as a present value discounted back to $t_I$, the introduction date for the new technology (fixed costs of diffusion are incurred instantaneously). The externalities generated depend on the time of diffusion. We assume that $t_D$ is such that it maximizes $E(t_D; t_I)$ with respect to $t_D$; i.e.,

$$\max_{t_D} E(t_D; t_I) = E[t_D(t_I), t_I]. \quad (12)$$

Henceforth we shall use the short notation $t_D^* = t_D(t_I)$ and $E^* = E[t_D(t_I), t_I]$.

The expression $E^*$ gives the magnitude of the *potential* external effects of introducing the new technology at $t_I$ discounted back to $t_I$. The optimum date of diffusion, $t_D^*$, is determined by the condition

$$\frac{\partial E(t_D; t_I)}{\partial t_D} = -\exp[-\rho(t_D - t_I)] \left\{ [\pi_2(t_D) - \bar{\pi}_2] \right.$$
$$\left. - \left[ \rho M_D(t_D; t_I) - \frac{\partial M_D}{\partial t_D} \right] \right\} = 0. \quad (13)$$

The term $[\pi_2(t_D) - \bar{\pi}_2]$ is the marginal *loss* from a delay in diffusion, while

$$G(t_D, t_I) \equiv \rho M_D(t_D, t_I) - \partial M_D/\partial t_D \quad (14)$$

is the marginal *gain*. The latter is composed of two elements: (*a*) the one-period service charge on the fixed diffusion costs, $\rho M_D$, which is saved if diffusion is delayed by one period; (*b*) the reduction in these fixed costs resulting from a marginal delay ($-\partial M_D/\partial t_D$). At $t_D^*$, the marginal loss and the marginal gain from a delay in diffusion must be the same (see fig. 11.2).[12]

---

11. With a separate owner of the new technology (as in Barzel's analysis), this requires charging users a royalty rate equal to the reduction in average production costs.

12. By specifying eq. (10) we obtain $G(t_D, t_I) = (1 + \rho)k/(t_D - t_I)$, from which it follows that

$$\lim_{t_D \to t_I} G = \infty; \quad \lim_{t_D \to \infty} G = 0.$$

The partial derivatives of $G$ with respect to $t_D$, $t_I$ are $G_D$, $G_I$, respectively, and

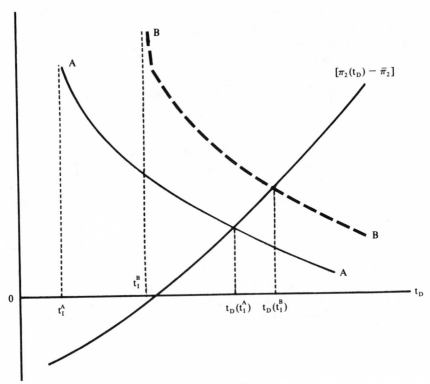

Fig. 11.2. Determination of optimum diffusion time, $t_D^*$.

AA:  $G(t_D, t_I^A)$.        BB:  $G(t_D, t_I^B)$.

### The effects of a delay in introduction

Differentiating equation (13) totally with respect to $t_I$ gives

$$t_D^{*'} = \frac{G_I}{\pi_2' - G_D}, \tag{15}$$

where $G_i \equiv \partial G/\partial t_i$ $(i = I, D)$. The denominator of this expression is positive (see note 12); it means that a small delay in diffusion will increase the marginal loss from a delay, $G$. The numerator represents the effect of a delay in introduction on the marginal gain from a delay in diffusion. It will always be positive (see note 12). Therefore, $1 > t_D^{*'} > 0$.

---

$$G_D = [-M_D/(t_D - t_I)][\rho + 2/(t_D - t_I)] = -G_I < 0.$$

Note that $\pi_2(t_D) - \bar{\pi}_2$ may be negative for low values of $t_D$. The second-order condition for $t_D^*$ to be optimal is $\pi_2'(t_D^*) - G_D > 0$. This holds since $\pi_2' > 0$ (see [A3]) and $G_D < 0$.

It is of interest to analyze the effects of an introduction delay on potential externalities discounted back to a fixed point of time, say $t = 0$, i.e., $\exp(-\rho t_I)E^*$ ($E^*$ is defined in eq. [12]). We first differentiate this expression totally with respect to $t_I$,

$$\frac{d}{dt_I}\exp(-\rho t_I)E^* = \exp(-\rho t_I)\left(\frac{dE^*}{dt_I} - \rho E^*\right). \quad (16)$$

It follows from equations (12)–(13) that the total and partial derivatives of $E^*$ with respect to $t_I$ are equal. Furthermore,

$$\frac{\partial E^*}{\partial t_I} - \rho E^* = -\exp[-\rho(t_D^* - t_I)]\frac{\partial M_D}{\partial t_I} \quad (17)$$

(see eq. [11]). Equation (16) can therefore be rewritten as

$$\frac{d}{dt_I}[-\exp(-\rho t_I)E^*] = \exp(-\rho t_D^*)\frac{\partial M_D}{\partial t_I} > 0. \quad (18)$$

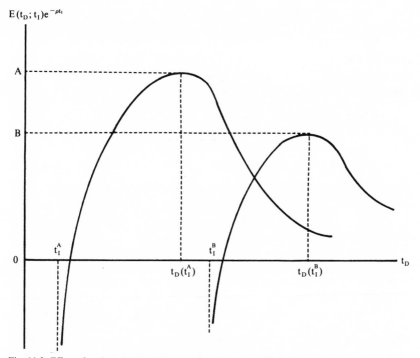

Fig. 11.3. Effect of an introduction delay on potential externalities, discounted back to $t = 0$.

$A = E^{*A}\exp(-\rho t_I^A)$.

$B = E^{*B}\exp(-\rho t_I^B)$.

The expression on the right-hand side of equation (18) may be termed the marginal externality gain from accelerating introduction, discounted back to $t = 0$. This result is depicted in figure 11.3.

The results of this section can be summarized in the following propositions:

> *Result 3*: A delay in introducing new technology into the economy will induce a delay in its diffusion.
>
> *Result 4*: A delay in introducing the new technology will increase the fixed costs of diffusion and will thereby reduce potential external economies discounted to the date when the technology became available.

## The Socially Optimal Introduction Date

Let $V_2$ $(t_I)$ be the present value at $t = 0$ of sector 2 profits when the new technology is introduced at $t_I$ and diffused at $t_D$ $(t_I)$;

$$V_2(t_I) = \int_0^{t_D^*} \bar{\pi}_2 \exp(-\rho\theta)\, d\theta + \int_{t_D^*}^{\infty} \pi_2(\theta) \exp(-\rho\theta) d\theta$$

$$- M_D(t_D^*; t_I) \exp(-\rho t_D^*)$$

$$= E^* \exp(-\rho t_I) + (\bar{\pi}_2/\rho) \tag{19}$$

(see equation [11]).

The socially optimal introduction date, $t_s$, will be that $t_I$ which maximizes $W(t_I) = V_1(t_I) + V_2(t_I)$. The first-order condition is

$$W'(t_s) = V_1'(t_s) + V_2'(t_s)$$

$$= \exp(-\rho t_s)\{\rho M_I - [\pi(t_s) - \bar{\pi}_1]\} + \frac{d}{dt_I} \exp(-\rho t_s) E^*$$

$$= 0 \tag{20}$$

from eq. [6]. The $t_I$ satisfying this condition is $t_s$. Rearranging, we get

$$\rho M_I = [\pi(t_s) - \bar{\pi}_1] + X(t_s). \tag{21}$$

Thus from equations (18) and (21),

$$X(t_I) = \exp(\rho t_I) \frac{d}{dt_I} \exp(-\rho t_I) E^*$$

$$= \exp[-\rho(t_D^* - t_I)] \frac{\partial M_D}{\partial t_T}. \tag{22}$$

The quantity $X(T_I)$ is the "marginal externality gain" from accelerating introduction, discounted back to $t_I$. The right-hand side of equation (21) is the marginal social gain of accelerating the introduction of the new technology. It is composed of the increase in sector 1 operating profits $[\pi_1$ $(t_s) - \bar{\pi}_1]$ and the

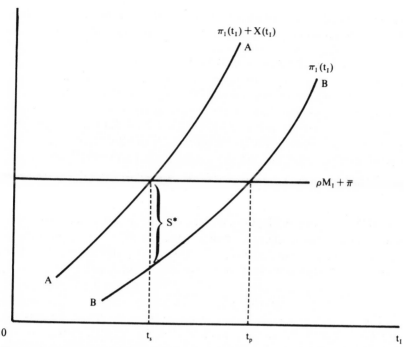

Fig. 11.4. Determination of $t_s$, $t_p$, and $S^*$.

marginal externality gain $X(t_s)$. The left-hand side is the marginal social loss, i.e., the one-period service charge on fixed introduction costs, $M_I$. At $t_s$ the marginal social gain should equal the marginal social loss. A graphical representation of the optimum social introduction time $t_s$ is shown in figure 11.4. The intersection between the $\pi_1(t_s)$ and the $\rho M_I + \bar{\pi}_1$ schedules gives us $t_p$, the optimum private introduction time. The intersection of the schedule for $\pi_1(t_I) + X(t_I)$ (schedule AA) and $\rho M_I + \bar{\pi}_1$ gives us $t_s$. Note that $t_p > t_s$.

### Technology-introduction subsidies

The producer of sector 1 may be induced to introduce the new technology at the socially optimal date by offering him a subsidy $S^*$ at a rate equal to the marginal externality gain from accelerating introduction at $t_s$. Thus, from equation (22),

$$S^* = X(t_s) = \exp[-\rho(t_D^* - t_s)] \frac{\partial M_D}{\partial t_I}, \qquad (23)$$

where $S^*$ is the optimum technology-introduction subsidy. From eqs. [22] and

[7], we can express $S^*$ as $S^* = \pi_1(t_p) - \pi_1(t_s) = (\rho M_I + \bar{\pi}_1) - \pi_1(t_s)$. Dividing both sides by $\rho M_I$, we get $S/M_I = 1 - [\pi_1(t_s) - \bar{\pi}_1]/\rho M_I$.

## Effects of Changes in Exogenous Variables

It is now possible to analyze the factors affecting dependent variables $t_p$, $t_s$, the $(t_p - t_s)$ gap, and the optimum subsidy, $S^*$. The effects of the following will be considered: increased profitability of the conventional technology and the new technology, and increased fixed costs of introducing and diffusing the new technology. The effects of having the profitability of conventional technology dependent on calendar time will be analyzed in the next section.

Before proceeding, it is useful to separate the exogenous factors pertaining to sector 2 [$\bar{\pi}_2$, $\pi_2(t)$, and $k$] from those pertaining to sector 1 [$\bar{\pi}_1$, $\pi_1(t)$, and $M_I$]. Changing levels of the former are associated with changing $X(t_I)$, the level of the marginal externality gain for *each* $t_I$. The $X(t_I)$ should be traced before considering the effects on the dependent variables. On the other hand, changes in sector 1 variables, while not affecting $X(t_I)$ directly, have an effect on the dependent variables that depends on the sign of $X'(t_I)$.

### The $X(t_I; Y)$ function

Let $Y$ be a sector 2 variable. From equation (13) and the specific formulation (10) it is clear that changes in $Y$ will affect the $t_D^*$ associated with a given $t_I$. Therefore $\partial X/\partial Y$ can be obtained from equation (22) by collecting the $\partial t_D/\partial Y$ terms:

$$\frac{\partial X}{\partial Y}\bigg|_{t_I} = \exp[-\rho(t_D^* - t_I)] \left( \frac{\partial M_D}{\partial t_I \partial t_D} \frac{\partial t_D^*}{\partial Y} - \rho \frac{\partial M_D}{\partial t_I} \frac{\partial t_D^*}{\partial Y} \right)$$

$$= \exp[-\rho(t_D^* - t_I)] G_I \frac{\partial t_D^*}{\partial Y} \sim - \frac{\partial t_D^*}{\partial Y} . \tag{24}$$

Thus a rise in $\bar{\pi}_2$ and $k$, by increasing $t_D^*$, will shift the $X(t_I)$ schedule downward. On the other hand, a rise in $\pi_2(t)$ for each $t$ will shift $X(t_I)$ upward.[13]

> *Result 5*: The partial effect of a change in an exogenous variable on the marginal externality gain will be in the opposite direction to its partial effect on the diffusion time of the new technology.

### The sign of $X'(t_I)$

Let us now consider whether the marginal externality gain $X$ increases or decreases with $t_I$. By differentiating equation (19) totally with respect to $t_I$, we get

---

13. If we substitute eq. (10) into eq. (13), we obtain an equation defining $t_D^*$: $\pi_2(t_D^*) - \bar{\pi}_2 = (1 + \rho)k/(t_D - t_I)$; and it follows that $\partial t_D^*/\partial \bar{\pi}_2 > 0$, $\partial t_D^*/\partial k > 0$, and $\partial t_D^*/\partial \pi_2(t) < 0$.

$$\frac{dX}{dt_I} = \frac{d}{dt_I} \exp[-\rho(t_D^* - t_I)] \frac{\partial M_D}{\partial t_I}$$

$$= \exp[-\rho(t_D^* - t_I)] \left[ \left( \frac{\partial^2 M_D}{\partial t_I^2} + \frac{\partial^2 M_D}{\partial t_I \partial t_D} t_D^{*'} \right) \right.$$

$$\left. - \rho \frac{\partial M_D}{\partial t_I} (t_D^{*'} - 1) \right]. \tag{25}$$

Recalling that $\partial^2 M_D /(\partial t_I^2 = 2 M_D / t_D^* - t_I)^2$ and $\partial M_D / \partial t_I = M_D /(t_D^* - t_I)$, making use of $G_I = -G_D$ (see note 12 and eqs. [14]–[15]), and collecting terms, we arrive at)

$$\frac{dX}{dt_I} = \exp[-\rho(t_D^* - t_I)] \frac{\pi_2' G_I}{\pi_2' + G_I} \sim \pi_2'(t_2^*) > 0. \tag{26}$$

*Result 6*: A delay in introducing the new technology into the economy results in an increase in the marginal externality gain. This is caused by the higher profitability of the new technology in sector 2.

This result explains why the slope of AA in figure 11.4 was drawn steeper than the slope of BB $[d(\pi_1 + X)/dt_I > d\pi_1 /dt_I > 0]$.[14] It will also explain some of the specific comparative-static results of the next section.

### Comparative-statics results

An increase in $\bar{\pi}_1$ and $M_I$ and a decrease in $\pi_1$ imply an increase in the relative advantage of the old over the new technology in sector 1, and similarly in sector 2 $(\bar{\pi}_2, k; \pi_2)$. The implications for $t_s$ and $S^*$ are shown in table 11.1 (they are obtained by simple shifts of the various schedules of fig. 11.4, taking account of eq. [21] and note 13). The major results that emerge can be summarized as follows:

*Result 7*: The greater the advantage of the conventional over the new technology in one or both sectors, the farther in the future will be the socially optimal introduction date for the new technology.

*Result 8*: The greater the advantage of the conventional over the new technology in sector 1, the greater will be the optimum technology-introduction subsidy; and the greater the advantage of the conventional over the new technology in sector 2, the smaller will be the optimum technology-introduction subsidy.

Result 7 is in accordance with what we would expect. Thus an increase in

---

14. This also ensures that $W(t_s)$ is a maximum. A sufficient condition is that $d(\pi_1 + X)/dt_I > 0$, which means that $dX/dt_I$ may be negative.

Table 11.1. The implications of an increase in the level of exogenous variables

|  | Effect on $t_s$ | Effect on $S^*$ |
|---|---|---|
| *Sector 1* | | |
| Conventional technology | | |
| Increase in $\bar{\pi}_1$ | + | + |
| Increase in $M_I$ | + | + |
| New technology | | |
| Increase in $\pi_1$ (for all $t$) | − | − |
| *Sector 2* | | |
| Conventional technology | | |
| Increase in $\bar{\pi}_2$ | + | − |
| Increase in $k$ | + | − |
| New technology | | |
| Increase in $\pi_2$ (for all $t$) | − | + |

the level of $\bar{\pi}_1$, $M_I$; $\bar{\pi}_2$, $k$ and a reduction in the level of $\pi_1$ ($t$) and $\pi_2$ ($t$) for all $t$, the greater will be $t_s$. Result 8 is partly obvious and partly not. In effect it states that $S^*$ and $t_s$ move in the same direction when the exogenous change pertains to the introducing sector (sector 1) and in opposite directions when it pertains to the diffusing sector (sector 2). How can we explain this?

The clue is to recognize that exogenous changes pertaining to sector 1 (sector 2) do not (do) directly affect $X$, the marginal externality gains from accelerating introduction. Thus changes in sector 1 which raise $t_s$ by virtue of $X'$ ($t_I$) $> 0$ (see eq. [26]), will necessarily increase $S^*$ [$= X(t_s)$]; that is, a relative improvement in the conventional technology of the sector, while delaying the socially optimal introduction date, will increase the optimum level of the technology-introduction subsidy. On the other hand, exogenous changes in sector 2 which raise $t_s$ (e.g., higher $\bar{\pi}$, $k$) *reduce* $X$ directly for each $t_I$. Moreover, this direct effect dominates the positive effect on $X$ induced by the higher $t_s$. Figure 11.5 shows the curves drawn in figure 11.4 plus an A'A' schedule reflecting the direct reduction of $X$ (downward shift in AA) resulting from the increased $\bar{\pi}_2$ or $k$. The original optimum subsidy is $S^* = CD$, while the new optimum subsidy is $S^* = \hat{C}\hat{D}$. Note that

$$\hat{S}^* - S^* = \hat{C}\hat{D} - CD = \hat{D}E = (F\hat{C} - \hat{D}E) - F\hat{C}. \qquad (27)$$

The quantity $\hat{C}\hat{D}$ is the partial increase in the optimum subsidy due to the increase from $t_s$ to $\hat{t}_s$. The quantity CD is the decrease due to the direct effect of the exogenous change. It clearly dominates the first term, so the effects on $t_s$ and on $S^*$ are in opposite directions.[15]

15. A similar analysis holds for the effects of a lower $\pi_2$ ($t$) (for all $t$).

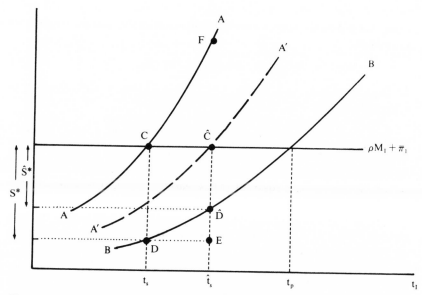

Fig. 11.5. Effects of an increase in $\bar{\pi}_2$, $k$, and a reduction in $\pi_2$.

## Accelerating the Availability of the New Technology

We now extend the framework to allow the date at which the new technology becomes available, $t_A$, to vary. This may be the result of shifts in the structure of government support of basic research and experimental development. As will be seen, delay or acceleration may have a marked effect on the introduction of the new technology when the profitability of the conventional technology is changing through time. We use the following notation:

$$t = t_A + \tau, \text{ calendar time (date);}$$
$$t_A = \text{availability date of the new technology;}$$
$$\tau = t - t_A, \text{ time elapsed } after \text{ availability;}$$
$$\tau_I = t_I - t_A, \text{ introduction delay;}$$
$$\tau_D = t_D - t_D, \text{ diffusion delay.}$$

We make the following specific assumptions: (B1) The profitability of current technology in the introducing sector, sector 1, is a monotonic function of calendar time,

$$\bar{\pi}_1 = \bar{\pi}_1(t) = \bar{\pi}_1(t_A + \tau), \quad \text{where } \bar{\pi}_1' \neq 0. \tag{28}$$

(B3) The profitability of the new technology in both sectors depends exclusively on the period elapsed after availability (this implies that assumption A9 still stands):

$$\pi_i = \pi_i(\tau), \quad \text{where } i = 1, 2 \text{ and } \pi_i' > 0. \tag{29}$$

We can now express the dependence of $V_1$, $V_2$, and $W$ on $t_A$ (all discounted to $t_A$) as follows:

$$W = V_1(\tau_I^*, t_A) + V_2(\tau_D^*, \tau_I^*) = W(t_A, \tau_I^*, \tau_D^*), \tag{30}$$

where $\tau_I^*$ may be either $\tau_s = \tau_s(t_A)$ or $\tau_p = \tau_p(t_A)$, and where $\tau_D^* = \tau_D(\tau_I)$. The social delay $\tau_s$ is obtained by setting $\partial W/\partial \tau_I = 0$, while $\tau_p$ is obtained by setting $\partial V_1/\partial \tau_I = 0$ (see eqs. [20], [6]). Clearly, $\tau_s$ and $\tau_p$ depend on $t_A$.

The quantity $V_2$ has the same form as in equation (19), with $\tau_D^*$, $\tau_I^*$ replacing $t_D^*$, $t_I^*$, respectively. The quantity $\tau_D^*$ has the same form as $t_D^*$ in equation (12). Thus $\tau_I^*$ is the optimum private or social introduction delay, given $t_A$. The quantity $V_1$ is given by

$$V_1 = \int_0^{\tau_I^*} \bar{\pi}_1(t) \, e^{-\rho \tau} d\tau + \int_{\tau_I^*}^{\infty} \pi_1(\tau) \, e^{-\rho \tau} d\tau - M_I e^{-\rho \tau} \tag{31}$$

(recall that $t = t_A + \tau$). $V$ now explicitly depends on $t_A$, both directly and through its effect on $\tau_I$ (cf. eqs. [3]–[5] above).

The basic issue analyzed here is the value to society of accelerating the availability of a new technology. Let $Z(t_A)$ be the present value (discounted to $t_A$) of the benefits of the new technology:

$$Z(t_A) = W - \bar{W}, \tag{32}$$

where

$$\bar{W} = \int_0^{\infty} \bar{\pi}_1(t_A + \tau) e^{-\rho \tau} d\tau + \bar{\pi}_2/\rho$$

is the present value (discounted to $t_A$) of the profits of sectors 1 and 2 under conventional technology. From equation (32) we obtain

$$Z'(t_A) = \frac{dW}{dt_A} - \frac{d\bar{W}}{dt_A}. \tag{33}$$

We now proceed to determine the sign of equation (33). The effect of a change in the availability date on $W$, the present value of the profits of sectors 1 and 2 discounted to $t_A$, is given by

$$\frac{dW}{dt_A} = \left( \frac{\partial V_1}{\partial t_A} + \frac{\partial V_1}{\partial \tau_I} \tau_I^{*\prime} \right) + \left( \frac{\partial V_2}{\partial \tau_I} \tau_I^{*\prime} + \frac{\partial V_2}{\partial \tau_D} \tau_D^{*\prime} \tau_I^{*\prime} \right). \tag{34}$$

This can be simplified by recalling that $\partial V_2/\partial \tau D = 0$ (see eqs. [19], [13]).

Evaluating equation (34) at $\tau_I = \tau_s$ and $\tau_I = \tau_p$, we get

$$\frac{dW}{dt_A} = \frac{\partial V_1}{\partial t_A} \quad \text{for } \tau_s, \tag{35a}$$

$$\frac{\partial W}{\partial t_A} = \frac{\partial V_1}{\partial t_A} + \frac{\partial V_2}{\partial \tau_I} \tau_p^{*\prime} \quad \text{for } \tau_p. \tag{35b}$$

Finally, substituting equations (35a) and (35b) into equation (34) and noting that $d\bar{W}/dt_A = \bar{\pi}_1' \exp(-\rho\tau) \, d\tau \sim -\bar{\pi}_1'$, we obtain

$$Z'(t_A)\big|_{\tau_s} = -\int_{\tau_s}^{\infty} \bar{\pi}_1 \, e^{-\rho\tau} d\tau \sim -\bar{\pi}_1', \tag{36a}$$

$$Z'(t_A)\big|_{\tau_p} = -\int_{\tau_p}^{\infty} \bar{\pi}_1 \, e^{-\rho\tau} d\tau + \frac{\partial V_2}{\partial \tau_I}\bigg|_{\tau_p} \tau_p'. \tag{36b}$$

Note that a delay in making the new technology available increases (decreases) the private introduction delay according as calendar time increases (decreases) the profitability of the conventional technology, i.e., $\tau_p^* \sim \bar{\pi}_1'$ (see fig. 11.6). Moreover, a reduction in the introduction delay, $\tau_I$, will necessarily increase $V_2$, i.e., $(\partial V_2/\partial \tau_I)_{\tau_p} < 0$ (since $\tau_p > \tau_s$).

Result 9: When availability is accelerated, the present value to society of the new technology (discounted to $t_A$) increases if the profitability of the conventional technology increases over time. It decreases if the profitability of conventional technology decreases through time. This holds for both the private and the social optimum introduction dates.

The reason why accelerating the availability of new technology may be critical when $\bar{\pi}_1' > 0$ can be shown with the aid of figure 11.6. The acceleration significantly reduces the introduction delay for the new technology; i.e., it increases the number of periods for which it will be generating benefits to society.[16] Under certain assumptions about endogenous changes in the operating efficiency of technologies (e.g., learning by doing), the early availability of the new technology may be critical for its effective introduction and diffusion throughout the economy.

### Summary and Conclusions

The main object of the paper has been to present a model of the introduction of a major new technology and its diffusion, as a framework for analysis of the role of government policy in the process. The main thrust of the analysis is

---

16. It will also reduce the optimum technology-introduction subsidy.

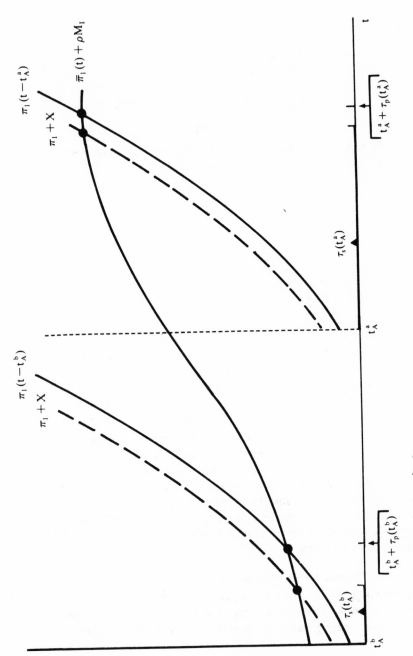

Fig. 11.6. Effects of reducing $t_A$ when $\bar{\pi}' > 0$.

the determination of the socially optimal introduction date, the optimum technology-introduction subsidy, and the effects of accelerating the availability of the new technology. We explicitly consider the experience generated by the introduction of the new technology—through sector 1—which makes possible its subsequent diffusion to sector 2. We assume that this experience creates unpriced benefits (externalities) to the latter sector.

The main conclusions can be summarized as follows:

1. The externalities generated by a new technology introduced at a particular date depend on the date of diffusion. There exists, for any introduction date, a date of diffusion which maximizes the value of the externalities. This maximum is termed the potential externalities from introduction at that particular date. Ceteris paribus, potential externalities decline when introduction is postponed.[17]

2. The greater the efficiency of conventional technology (in either sector), the longer the delay (both private and social) in introducing the new technology. Moreover, the optimum subsidy required increases with the efficiency of the introducing sector (1) and decreases as the efficiency of the diffusing sector (2) increases.

3. The notion of "marginal externality gain" from accelerating the introduction of a new technology is useful in analyzing the effects of exogenous changes on the socially optimal introduction date and on the optimum subsidy, $S^*$. When an exogenous change (e.g., a change in the efficiency of the old or new technology in the diffusing sector or a change in the diffusion costs of the new technology) affects the marginal externality gain directly, a delay in $t_s$ will be associated with a reduction in $S^*$ and vice versa. When an exogenous change (such as a change in efficiency in the introducing sector or a change in the fixed introduction costs of the new technology) does not have a direct effect on the marginal externality gain, a delay in $t_s$ will be accompanied by an increase in $S^*$ and vice versa.

4. Accelerating the availability of a major new technology may create substantial economic benefits, especially when the efficiency of conventional technology in the sector of introduction is increasing through time. In some cases, it may even be critical for the introduction of the new technology and therefore the possibility of its diffusion to a wide spectrum of economic activities. In other cases, a small acceleration in availability may reduce the introduction delay substantially.[18]

17. One implication of the existence of externalities is that, up to a point, competition in introduction reduces the gap between $t_p$ and $t_s$ for a *major* new technology. This is not true of other innovations (see Barzel 1968; Kamien and Schwartz 1972).

18. This issue has also been analyzed by Utterback and Abernathy (1975) and Teubal (1979), who deal with the lack of innovation incentives in the mature stage of the cycle of an industry or

5. A first glance at the government's role in this area shows that support for availability or introduction (or both) of new technologies is relatively widespread. However, the reason for this support seems to be the satisfaction of some extraeconomic national goal (e.g., defense or space exploration) rather than the perceived economic desirability of advancing the date of availability or of taking externalities into account. Examples are computers, integrated circuits, and recent technologies in the electronic industry.[19]

The model requires further development and refinement in order to provide a useful framework for decision making (in, for example, issues relating to energy). In particular, endogenous changes in the efficiency of the conventional technology spurred by the recent availability or introduction of a new technology have not been considered; nor has the choice between alternative new technologies been considered.

The paper as it stands, however, raises some issues concerning government policy directed at maintaining an aggregate rate of economic growth when the growth rate of individual sectors is observed to slow down (see Kuznets 1930 and Burns 1934). Attainment of this objective requires the introduction of new technologies, a phenomenon that is bound to be associated with externalities, including those of the type considered in this paper. It is our belief that an important aspect of macroeconomic theory and policy should deal with the search for and evaluation of such technologies, the strategies for their development and implementation, the timing of their introduction and diffusion, and the associated institutional arrangements. The energy crisis is bound to make us, as economists, more aware of these needs.

### References

Arrow, Kenneth J. 1962. "Economic Welfare and the Allocation of Resources for Invention." In *The Rate and Direction of Inventive Activity: Economic and Social Factors*, edited by R. R. Nelson, 609–25. Princeton: Princeton University Press.

---

product class, the former concentrating on the supply and the latter on the demand factors involved. A possible example in the real world is the automobile industry in relation to electric cars since the Second World War. Another possible example is in communications: the efficiency of cable and microwave relay systems played a role in determining that satellite communications technology was introduced in international, rather than United States, domestic communications. In the electronics industry the efficiency level and improvements in random access memory devices (RAM) are creating obstacles to the diffusion of novel new memory technologies such as charge-coupled devices. See Gosney 1979.

19. Acceleration of availability or introduction of a major new technology for noneconomic reasons need not eliminate the desirability of accelerating because of externalities or other economic benefits. It may conceivably justify an even greater effort in that direction. Further thought on this point is called for.

Barzel, Yoram. 1968. "Optimal Timing of Innovations." *Review of Economics and Statistics* 50 (August): 348–55.

Burns, Arthur F. 1934. *Production Trends in the United States since* 1870. National Bureau of Economic Research; General Series no. 23. New York: NBER.

Freeman, Christopher. 1974. *The Economics of Industrial Innovation.* Harmondsworth: Penguin Books.

Gosney, Milton. 1979. "Reappraising CCD Memories: Can They Stand up to RAM's?" *Electronics International* 7 (June): 122–26.

Kamien, Morton I., and Nancy L. Schwartz. 1972. "Timing of Innovations under Rivalry." *Econometrica* 40 (January): 43–60.

Kuznets, Simon. 1930. *Secular Movements in Production and Prices.* Clifton, N.J.: Kelley.

Kuznets, Simon. 1971. *Economic Growth of Nations, Total Output and Production Structure.* Cambridge: Harvard University Press.

Kuznets, Simon. 1975. "Technological Innovation and Economic Growth." In *Technological Innovation: A Critical Review of Current Knowledge*, edited by P. Kelly and M. Kranzberg. San Francisco: San Francisco Press.

Mathematica, Inc. 1975. *Quantifying the Benefits to the National Economy for Secondary Applications of NASA Technology.* Report prepared for NASA. Princeton, N.J.

Midwest Research Institute. 1971. *Economic Impact of Stimulated Technological Activity.* Part 2: *Case Study— Technological Progress and Commercialization of Communications Satellites.* Report prepared for NASA.

Robbins, Martin D., John A. Kelley, and Linda Elliott. 1972. *Mission-oriented R & D and the Advancement of Technology: The Impact of NASA Contributions.* Report prepared for NASA. Denver: Denver Research Institute, University of Denver.

Rosenberg, Nathan. 1963. "Technological Change in the U.S. Machine-Tool Industry, 1840–1910." *Journal of Economic History* 23 (December): 414–43. Reprinted in *Perspectives on Technology*, chap. 1. London and New York: Cambridge University Press, 1976.

Scherer, Frederic M. 1967. "Research and Development Resource Allocation under Rivalry." *Quarterly Journal of Economics* 81 (August): 359–94.

Teubal, Morris. 1979. "Need Determination, Product Type and the Inducements to Innovate." Discussion Paper no. 791. Maurice Falk Institute for Economic Research in Israel, Jerusalem.

Utterback, James M., and W. J. Abernathy. 1975. "A Dynamic Model of Process and Product Innovation." *Omega* 3 (6): 639–56.

A preliminary version of this paper appeared as part 3 of the project entitled "Governrment Policy, Innovation, and Economic Growth: Lessons from a Study of Satellite Communications." The project was financed by the NASA-Ames Research Center, Moffett Field, California, under Interchange no. NCA2-OR745-815. We are grateful to Susanne Freund for her comments and suggestions, which considerably improved preliminary drafts of this paper.

# 12 The Engineering Sector in a Model of Economic Development

## Morris Teubal

The purpose of this chapter is to analyze some issues related to the role of the engineering (capital goods) sector in transmitting growth from an expanding primary-goods export sector to the rest of the economy. One possibility explored in a previous paper (Teubal 1980) is to divide the economy into three sectors: the primary sector, sector X, whose growth in response to a growing world market generated possibilities, directly and indirectly, for the emergence and growth of other sectors; the engineering sector, sector E, whose output of machines or intermediate inputs is directed toward satisfying the local demand created by the expanding primary sector (backward linkages) and where the production skill accumulated increases the effectiveness of resources invested in developing new techniques for the remaining sectors (spin-off);[1] the rest of the economy, sector C, whose existing techniques cannot sustain domestic production but where an appropriate technique may be developed, thanks to the skills being accumulated in the engineering sector.

The first section summarizes the objectives and structure of the model and presents some of the conclusions derived in the original paper (Teubal 1980). Next, a more thorough discussion of the nature of the model relative to other growth models and of specific issues emerging from it will be undertaken. Attention will be given to the following: (*i*) the initial stimulus to growth; (*ii*) the focus of the analysis; (*iii*) endogenous technical changes; (*iv*) the importance of chronological time; (*v*) the time sequence of activities; and (*vi*) policy issues.

---

1. In this chapter, the expression "engineering sector" has the same meaning as the term "capital-goods sector" or "machinery-producing sector."

---

## The Model

### Objectives

The first objective of the model is to formalize some of the roles that economic historians and development economists have assigned to the metalworking/machine-building sectors. It has been suggested that in this sector the skills (components) acquired (developed) while performing certain activities or tasks can then be fruitfully applied to other activities or tasks.[2] In our model, production skills help design activities, the former activity aimed at satisfying input demands from the primary sectors and the latter activity aimed at enabling the substitution of imports of other goods for domestic production.

A second objective of the model is to incorporate such a sector into a multisector model of the economy where some initial stimulus sets the growth process going in one sector and where a series of factors and parameters will determine whether or not the growth is transmitted to other sectors. The engineering sector will play a crucial role in this transmission.[3] In terms of our model, we ask ourselves under what conditions will a C sector emerge in our economy, the initial bottleneck being the absence of a technique on the shelf that is appropriate to the conditions of our economy.

### Description

A very short description will be presented here. Further details can be found in Teubal 1980.

#### Primary Sector

It is assumed that our economy may influence the world price of its primary export and that world export demand grows at an exponential rate $g$ at least until chronological time $T$ when, under one interpretation of the model, it

---

2. Rosenberg (1963) uses the term "key sector" when describing the activities of the United States machine-tool sector in the nineteenth century. The sector developed and diffused technologies for a wide spectrum of metal-using, metal-forming activities. Diffusion involved the adaptation of a machine tool already developed for one sector to the specific needs of another. Any single adaptation enlarged the pool of skills and knowledge relevant for subsequent adaptations. Pack and Todaro (1969) and others refer to the role of the machine sector in adapting foreign technology to the local needs. Youngson (1959) refers to the role of machinery and implements firms in developing new techniques in response to changing world market conditions. For a survey of part of the literature, see Stewart (1978).

3. Our model thus also gives a formal expression to the central issue discussed by the so-called staple theories of export-led growth; see Watkins (1963); Baldwin (1956). For a good review, see Roemer (1970).

collapses. This specific interpretation will be adopted here because it fits well with an aspect of the model that we want to emphasize, namely, that the exogenous stimulus to growth is a temporary one, like a short-lived staple (see below for further comment). Alternatively, $T$ may be considered as the planning horizon of entrepreneurs in the engineering sector, but this fits less well with our objective, and it requires modifying the model somewhat (the central issue would be the time of appearance of sector C rather than whether C will appear or not). It is also assumed that there exists perfect competition, surplus labor, and that the everlasting capital is acquired from abroad until $t_1$, when it begins to be supplied domestically (see the later discussion). Finally, production takes place under constant costs, and these remain constant through time.[4] Under these conditions, the growth of the primary sector X is represented by the equation

$$x_t = x_o e^{gt}, \tag{1}$$

where $x_t$ is primary sector output (exports) at time $t$.

### Engineering Sector

Sector E emerges at $t_1$ to supply inputs to the primary sector once the local market $y_k^d$ is sufficiently large to enable unit production costs $c_k$ to be lower than the fixed world price $\bar{p}_k$. There are static production economies of scale

$$c_k = a + y_k^{-\alpha}. \tag{2}$$

where $y_k$ is the rate of output and $\alpha$ is a positive parameter. This function is invariant through time. Exports are not possible and local demand for the infinite-lived capital is given by

$$y_k^d = \mu_x \, dx/dt = \mu_x g x_0 e^{gt}, \tag{3}$$

where $\mu_x$ is a fixed-capital-output coefficient. It follows that $t_1$ is defined by the equation

$$K_x^M \equiv \mu_x x_0 \exp(g t_1) = 1/g(\bar{p}_k - a)^{1/\alpha}, \tag{4}$$

where $K_x^M$ is the stock of imported capital in the primary sector at and beyond $t_1$. We assume that the machines sold after $t_1$ to the primary sector are priced at $\bar{p}_k$, their import cost to $X$.[5]

---

4. Labor is being drawn at a constant wage $\bar{w}$, and the cost of capital to the sector is assumed to be constant through time. In particular, the shift to local production of capital goods is assumed not to be associated with a reduction in the price paid by the sector, which will be discussed later.

5. This is profit maximizing, provided the discontinuous drop of marginal revenue at $\bar{p}_k$ makes it attain a level below the level of marginal costs at that point. A sufficient condition is an inelastic demand at that point, although this is not necessary.

The sudden shift from imports to domestic production postulated in this model is an oversimplification of reality. It would be desirable to describe a gradual transition, but this would complicate the model without apparently modifying the nature of the results. Concerning exports of capital goods, it is reasonable to assume that nonprice factors and setup costs preclude exports at the early stage of this industry. The assumption that no exports are possible is, however, increasingly less realistic as time elapses beyond $t_1$ and the sector becomes more fully developed. In this paper, we concentrate on the emerging stages of the sector.

### Sector C

Sector C represents the rest of the economy and, for simplicity, is assumed to be a consumption-goods sector. Our country does not have a comparative advantage initially in producing these goods because the technique—identified here by the capital-labor ratio—which is available from the advanced countries is too capital-intensive and therefore inappropriate.[6] However, the production experience accumulated by sector E enhances its design capabilities in respect to the techniques for sector C. Specifically, the fixed design costs of developing technique $k_c$ at $t_2$ are given by the expression

$$d(k_c, t_2) = \beta(k_c^A - k_c)/V(t_2), \tag{5}$$

where $k_c^A$ is the C technique available on the shelf from advanced countries, $\beta$ is a positive parameter, and $V(t_2)$ is the design-relevant production experience accumulated in sector E,

$$V(t_2) = \left\{ \int_{t_1}^{t_2} y_k(t) \, dt \right\}^{\eta}, \tag{6}$$

$\eta$ translating production experience to design capabilities.

### Emergence of Sector C

Entrepreneurs make a cost-benefit analysis at $t_1$ to determine whether or not to introduce a technique for sector C, and if so, which $k_c$ and at what $t_2$. We assume that there exists a monopolist E firm that extracts all the "surplus" from potential C producers; i.e., unit royalties $R(k_c)$ are given by

$$R(k_c) = 1 - C_c(k_c),$$

where 1 is the world price of C goods—the *numéraire*—and $C_c(k_c)$ the unit production cost when technique $k_c$ is being used. Unit cost declines with $k_c$

---

6. Inappropriateness could also derive from other factors, such as intensiveness in natural resources, that are not abundant in our economy.

within an interval $(k_c^A, \underline{k}_c)$ and increases thereafter. Total royalties per period, until time $T$, the time of collapse of the primary sector, are discounted back to $t_1$ and compared with discounted, fixed development costs. Let

$$B(t_2, k_c) = R(k_c) \int_{t_2}^{T} y_c(t) \exp[-\rho(t - t_2)]dt \tag{7}$$

be accumulated royalties discounted back to $t_2$ at rate $\rho$, with $y_c(t)$ being the output of sector C at time $t$.

A technique will be developed for sector C enabling the sector to be established if and only if there exists a technique $k_c$ $(k_c^A > k_c > \underline{k}_c)$ and a time $t_2 > t_1$ such that

$$B(t_2, k_c) \geq d(t_2, k_c). \tag{8}$$

This condition can be elucidated by combining all the terms into two functions: $f(t_2, T)$, which includes all the $t_2$ terms, and $g(k_c)$:

$$f(t_2, T) = V(t_2) \int_{t_2}^{T} Y_x(t) \exp[-\rho(t - t_2)]dt, \tag{9}$$

$$g(k_c) = \beta(k_c^A - k_c)/m(k_c, \gamma) R(k_c), \tag{10}$$

where $Y_x(t)$ is the income of domestic factors in sector X, i.e., net of capital service charges paid forever on the stock of capital, and $m(k_c, \gamma)$ is a sort of "multiplier" relating $Y_x(t)$ to the local demand for sector C goods, with $\gamma$ the constant proportion of the income of domestic factors of all sectors spent on C goods. It depends, among other things, on the distribution of income.

The first function, $f(t_2, T)$, is the product of the domestic value-added of the primary sector from the date when the technique is developed $(t_2)$ until the collapse of the sector (at $T$) times the experience term $V$. Both elements favor the development of the new technique.

A rise in $t_2$ enables the accumulation of experience but it shortens the time left before primary exports collapse. The $f$ function will therefore be bell-shaped with respect to $t_2$.

A necessary condition for the emergence of sector C is

$$\max f(t_2, T) \geq \min g(k_c),$$

as shown in figure 12.1$a$. Figure 12.1$b$ depicts a situation where no technique will be developed by that sector, so no sector C will emerge prior to the collapse of X.

### Conclusions

Factors raising the $f(t_2, T)$ curve and lowering min $g(k_c)$ increase the probability of the emergence of sector C. It turns out that there are a number of strategic variables.

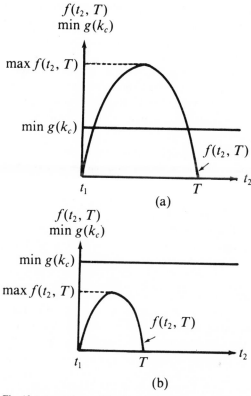

Fig. 12.1. Possible conditions for the (a) emergence and (b) nonemergence of sector C.

### Size of Local Market

The local market size is a main determinant of the profitability of introducing a technique into sector C because exporting is inherently more difficult than selling to the local market. The main variables affecting it are: the rate of growth of exports $g$, the proportion $\gamma$ spent on C goods, and the stock of imported capital in the export sector $K_x^M$. We shall deal with each one separately. For $g$, a higher rate of growth of exports implies a greater export sector income, a wider local market for C goods, and greater prospects for the emergence of this sector. For $\gamma$, its magnitude depends fundamentally on the distribution of income in the export sector. A more egalitarian distribution of income will, presumably, lead to a greater domestic market for C goods.[7] For

---

7. This is the general impression one obtains from the case study literature on exports and economic development. Compare the case of wheat in Canada with sugar cane, copper, etc. A useful review of this issue can be found in Roemer 1970.

$K_x^M$, a higher value implies greater capital service flows abroad and therefore a lower income received by domestic factors of production.

### Learning Elasticity $\eta$

The elasticity coefficient relates the relevant aspects of production experience to design costs. In a sense, it refers to a transfer of skills from one activity to the other. A certain hierarchy of skills (maintenance and repair, production, design) underlies the approach followed in this chapter.

### The Horizon T

There is a horizon $T$ beyond which the primary sector ceases to be viable. A higher $T$ shifts the $f(t_2, T)$ curve upward and to the right, thereby increasing the probability that a C technique will be profitable. This occurs for two reasons—one demand and one supply: (i) the primary sector will be earning income and spending it on C goods for a longer period of time; (ii) more design-relevant production experience can be accumulated in the engineering sector.

### Returns to Scale in Sector E

A higher level of $\alpha$ results in an earlier appearance of sector E. This enables more production experience to be accumulated in this sector throughout the lifetime of the primary sector, thereby increasing the profitability of developing a technique for C.[8]

## Nature and Implications of the Model

The three-sector model just presented is significantly different from the three-sector growth model I proposed some time ago (Teubal 1971, 1973) and from neoclassical growth models generally. Although some of the latter incorporate the possibility of technical change or learning by doing, the description of these phenomena is invariably too simple and even trivial. The effect of learning, for example, is to reduce the production cost of *existing products* (Bardhan 1970), a fact that is not the essential feature of sectors whose major characteristic is

---

8. It should be noted here that $\alpha$ determines the *elasticity* of unit costs $c_k$ with respect to output $y_k$, so a higher level of this variable, ceteris paribus, should be favorable to the developing country. Another crucial scale parameter not yet considered is the *level* of unit costs at a particular output level. A small firm contemplating producing such a good should be particularly interested in the level of unit costs at low levels of output, relative to the import price of the good $\bar{P}_k$. For a given elasticity of unit costs, the higher such a level—a variable associated with the *diseconomies of small scale*—the *less* favorable it will be for the emergence of sector E. The cost function postulated in this chapter does not permit parameter variation of this variable because unit costs at zero outputs always tend to infinity.

specialization in a specific technology, such as mechanical engineering or electronics, rather than in specific products. Such a feature is the capability of doing *new* things as spin-offs from current or past activity. In explicitly postulating such a sector, albeit in an extremely simple fashion, this model differs from other growth or development models.[9] Beyond this major characterization of the model, we find a series of characteristics that merit separate attention.

### Growth-Initiating Factor (Initial Stimulus)

In our model, the stimulus to growth derives from an expansion in the world demand for the economy's primary product (demand pull). No saving or capital accumulation is needed, even as an enabling factor, because we assume that capital can be rented from abroad. In contrast, most neoclassical models generally consider saving and capital accumulation as factors initiating growth (supply push).[10] In addition, by postulating the existence of a saving function and a constant rate of population growth, they imply that the stimulus to growth is a *permanent* one. A major feature of our model, on the other hand, is that the exogenous stimulus to growth is a *temporary* one: world demand, although first growing at an exponential rate, collapses for some unspecified reason at time *T*. Although there are other, more elegant, formulations for a temporary exogenous stimulus to growth, the main point remains that the environment of developing countries is subject to violent fluctuations. This implies that (*a*) it is not wise to rely on any particular exogenous growth stimulus for long and (*b*) the countries involved should develop capabilities for exploiting new growth opportunities once they arise.

### Focus of Analysis: Environment and Bottlenecks

Our analysis focuses on the conditions for the emergence of sector C, where the effective bottleneck is absence of an appropriate technique of production. Alternatively, we focus on the conditions assuring the *transmission of growth* from the primary sector. These conditions are expressed in terms of the following staple-theory-type *indirect* effects of export expansion: production linkage, consumption-goods linkage, skill, and innovation. Neoclassical models, on the other hand, tend to focus on the very long run *steady state*: its characteristics,

---

9. The fact that such sectors exist in reality also provides a justification for assuming that the learning and externality phenomena are concentrated in one or a small number of such sectors rather than presuming, out of ignorance, that they may be equally distributed among all sectors of the economy.

10. See the models of Solow (1956) and Uzawa (1964). Exceptions to this are the models of Bardhan (1970) and Findlay (1973, chap. 5). In these models, although the opportunity to grow derives from expansion in world demand (as in our model), the process also requires saving and capital accumulation.

such as rate of growth, and the conditions assuring convergence of the factor accumulation path to a balanced growth path.[11]

Our concern for the transmission of growth from the sector receiving the initial stimulus stems from the nature of the environment facing our less-developed economy. The issue arises as to what remains after the primary sector disappears or after its significance has been substantially reduced. This is a more fundamental issue than the issue of convergence toward, and nature of, the steady state.

### Endogenous Technical Change

Unlike other formal models of growth or development, the present one stresses the primacy of technological innovation for the emergence of new sectors (beyond X and E). This enables the model to capture what seems to be a central fact: a series of microstudies of the last few years have shown quite convincingly that success in the transfer of technology to developing countries requires local adaptation of technology.[12] In other words, an active recipient of the foreign technology is required in order to adapt foreign technology successfully to local conditions. This is probably even truer with respect to technologies enabling—as in our model—the establishment of a new sector. The absence of a local capability to adapt foreign technology would represent, under those circumstances, a skill, or know-how, bottleneck to the establishment of such a sector.[13]

The analysis focuses on the supply and demand factors determining whether or not the appropriate technique will be developed. Both sets of factors are present here: Supply is represented by the accretion of production experience relevant to design, that is, a learning-by-doing phenomenon. Demand is represented by the growing *local* demand for C goods, a consequence of growing export proceeds and a sharp differentiation between the local and the world markets—specifically, at the early stages of development of C, the country cannot export.

Concerning the relevance of the skill acquisition process postulated, the

---

11. A less common objective is describing structural change along the growth path of the economy; see Teubal (1971, 1973); and Findlay 1973.

12. See Katz 1976 and the series of studies on adaptation and local generation of technology undertaken under the ECLA/IDB/IDRC/UNDP Research Programme on Scientific and Technological Development in Latin America. See Teitel, 1981.

13. The commissioning abroad of the required adaptation, although possible in principle, may be impractical for a number of reasons, at least within the relevant time framework. The reasons include lack of intimate knowledge of local needs and the high price the developing country would have to pay.

importance of learning by doing has been consistently pointed out in the economic development literature. Lack of ability to invest in less-developed countries is ascribed to the absence of experience with a modern sector (Hirschman 1958). Cooper (1973) states that this type of skill creation is a basic ingredient to economic development and that market failure in providing enough opportunities, owing to externalities and risk, is a major cause of technological dependence. Leff's (1968) interpretation of growth and increasing sophistication of the Brazilian capital-goods industry is very much based on just such a type of skill acquisition process. Activity sequences, such as production first, design later, as postulated in the model, are also encountered in the literature, including a shift from repair and maintenance activities on machinery and equipment to production of machinery parts and components.[14]

The sharp distinction between the local and foreign markets postulated in the model is due not only to transportation costs but also to informational requirements on foreign markets, which demand time for acquisition; to the high setup (fixed) marketing and distribution costs associated with exporting; to obstacles caused by a firm's lack of reputation; and to low and nonhomogeneous quality of products at the early stages of the development of an industry. In this sense, a sharp difference probably exists between staple exports and exports of certain categories of industrial products.

### The Importance of Chronological Time

In neoclassical models, chronological time does not play a role when addressing the central issue dealt with: the nature of and convergence toward the steady state. These depend on the parameters of the model and are independent of the initial conditions of the economy—initial capital-labor ratio and (if desired) chronological time. In our model, chronological time is central because the exogenous stimulus to growth depends on this variable, and specifically, it terminates at time $T$. The structure of the model is such that the transmission of growth from the export sector should also take place before $T$. It follows that initial conditions, which, in our model, could be represented by the date of discovery of those natural resources that the primary sector will exploit, play a central role. If the "beginning" is late relative to $T$, then the chances that the development of the export sector will lead to the emergence of C are correspondingly low.

14. See Leff (1968) for examples within the Brazilian capital-goods industry and Roemer (1970) for examples such as that of copper mining in Zambia. Israel Aircraft Industries is a good example of a shift from repair of aircraft to production of parts and components, design of new components, and even of completely new aircraft. For an interesting link between preventive maintenance activities and design activities in an Argentine steel plant, see Maxwell 1976.

Sequence of Activities

The time sequence of sectors that may appear in our model is the following: first, the primary exporting sector (sector X), then the sector producing inputs for the primary sector (sector E), and finally, a manufacturing consumer-goods sector (sector C). This structure is particularly useful for analyzing the staple theory of export-led growth, namely, the conditions under which export growth will be transmitted to the rest of the economy. It would be presumptuous, however, to ascertain that this is the optimum time sequence of activities for *all* developing countries or that historical experience has always conformed to this pattern. The potential significance of the model does not lie in the specific sequence postulated but rather in having attempted a conceptualization of the engineering sector and its incorporation into a broader model of economic development. There are some historical examples where elements of the pattern suggested in this chapter seem to be present.

1. In his study of the capital-goods industries in Brazil, Leff (1968, 118–29) states explicitly that their emergence occurred very early in Brazil's development, in response to the processing needs of primary products such as coffee, sugar, and cotton. The overall impression is that at least the emergence of the Brazilian capital-goods industry is not the final stage of an industrialization process beginning with final consumer industries like textiles and food. The study also gives examples of spin-offs from the early activities of machinery-producing firms.

2. The development of Denmark and Sweden in the nineteenth century, as described by Youngson (1959), also includes aspects of machinery and implements firms developing in response to a rapidly expanding export sector (timber, dairy products, etc.). However, this development cannot be easily separated from the impulse derived from "the general development of industry," i.e., from manufacturing that was probably largely devoted, in that period, to satisfying local demand. This probably has also been the case in the economic history of other present-day developed countries.

However, despite these historical cases, there seems to be significant historical evidence favoring the "normal" patern, where the production of consumer goods precedes that of capital goods. The model, we emphasize again, should therefore be regarded as one possible sequence among a set of sequences, one that need not be the most frequent one. The sequence postulated is presumably more realistic for the subset of cases where the external stimulus to growth (via export demand) was significant relative to the stimulus derived from the local market.

A specific criticism is that the pattern of structural change postulated in the model contradicts the import substitution industrialization (ISI) experience of developing countries in Latin America after World War II, where consumer or final goods development preceded the development of intermediate and capital goods. This is undoubtedly true, but two questions arise:

1. The model presented is, at present, only a descriptive model; i.e., the time sequence of activities is that which appears in the absence of governmental policy. On the other hand, the ISI policy has been implemented by massive protection of the industrial sector as a whole. The historical record, therefore, need not contradict the results of the model. In particular, a tariff or production subsidy $\tau$ on C goods

$$\tau \geq c(k_c^A) - 1$$

imposed before $t_1$, the time of emergence of the E sector, would lead to the prior emergence of the C sector.

2. The ISI policy was at least a partial failure. Contradiction of the historical record may, in this case, represent an advantage of the model rather than a disadvantage. Moreover, by highlighting the difference between the engineering sector and other sectors, the model suggests the need for *selective* government support directed to this sector rather than all-out support of (other) manufacturing.

### Policy Issues

Some preliminary thoughts on the objectives and effects of policy are appropriate here in order to elicit comments. It is not clear what the objectives of policy should be because they depend on the decision makers' perceptions of the opportunities and risks to the economy beyond calendar time $T$. Although the descriptive part of this chapter focused on the emergence of C, the normative part may focus on attaining a sufficiently well developed engineering sector by $T$ to enable the economy to adapt to the new environment confronting it — all this at a minimum cost to the economy.

One aspect of this is the time of emergence of the engineering sector. There are no unpriced benefits (externalities) in our model because we assumed that the engineering sector was a monopoly. However, we assumed that the beginning of this sector is the result of a myopic decision; i.e., it takes place once local demand for inputs to X is sufficiently great to enable unit costs to equal the world price. The economic value of the benefits from production experience — the spin-offs — are not taken into account. Therefore, the policy implications are similar to the case of externalities: the emergence of the engineering sector should be anticipated relative to the date it would begin operations under market forces. This conclusion is relatively standard and does not depend on the terminal value of the skills accumulated by this sector until $T$.[15]

---

15. It depends on this terminal value being positive if no spin-off should take place prior to $T$.

In choosing the policy instruments to attain this objective, consideration should be given to the existence of differential learning effects. A production subsidy will unambiguously raise the production experience of sector E by

$$\int_{t_1'}^{t_1} y_k(t) \, dt,$$

that is, by the accumulated output of E goods between the anticipated time of emergence $t_1'$ and the original time $t_1$ ($t_1' < t_1$). The level of output beyond $t_1$ will remain as before.[16] An equivalent tariff on imports, on the other hand, has two opposite effects on production experience: first, a positive effect from the anticipation of domestic production of E goods; second, a negative effect from a substitution of capital goods by labor in the primary sector.[17] The former refers to the ($t_1'$, $t_1$) period and the latter takes place beyond $t_1$. The net effect on production experience is not known without further specification of the model. We would expect it to be generally lower than that resulting from an equivalent subsidy or even to be negative. In the latter situation, tariff protection of the engineering sector will *reduce* the spin-off that would result from this sector.

### Final Remarks

We have commented on a three-sector model of an economy with an engineering or machinery-producing sector. Differences and similarities with other formal models have been analyzed. The pattern of structural change and some policy issues have also been considered.

Additional work is required regarding the basic framework developed up to now and with respect to policy issues. The existence of uncertainty, especially with regard to the future world demand for primary exports, should be explicitly modeled. This would improve on the assumption made in this chapter that the primary sector collapses at a specified future date $T$. The processes of capital accumulation and the balance-of-payments constraint should also be introduced, and the appropriate set of policy measures favoring the engineering sector should be worked out. A continuous effort should be made to relate the assumptions, structure, and conclusions derived from the model to the

16. This occurs because the sector will continue to charge a price $\bar{p}_k$ to X producers, except when the marginal costs (associated with the quantity demanded at $\bar{p}_k$) *net* of subsidies declines below marginal revenue.

17. This will include both a pure substitution and an "output" effect owing to an increase in the costs of primary production.

historical case studies and to the growing body of findings on innovation and development. Needless to say, the task of conceptualizing and formalizing an engineering sector or firm is still in its beginnings.

## References

Baldwin, R. 1956. "Patterns of Development in Newly Settled Regions." *Manchester School of Economic and Social Studies* 24 (May): 161–79.

Bardhan, P. 1970. *Economic Growth, Development, and Foreign Trade.* New York: Wiley.

Cooper, C. 1973. "Science, Technology and Production in the Underdeveloped Countries: An Introduction." In *Science, Technology and Development,* edited by C. Cooper. London: Frank Cass.

Findlay, R. 1973. *International Trade and Development Theory.* New York: Columbia University Press.

Hirschman, A. 1958. *The Strategy of Economic Development.* New Haven: Yale University Press.

Katz, J. 1976. *Importación de tecnología, aprendizaje e industrialización dependiente.* Mexico: Fondo de Cultura Economica.

Leff, N. 1968. *The Brazilian Capital Goods Industry, 1929–1964.* Cambridge: Harvard University Press.

Maxwell, P. 1976. "Learning and Technical Change in the Steel Plant of Acindar." IDB/ECLA/UNDP Research Program on Science and Technology, Working Paper no. 4. Buenos Aires.

Pack, H., and M. Todaro. 1969. "Technological Change, Labor Absorption and Economic Development." *Oxford Economic Papers* 21 (November): 395–403.

Roemer, M. 1970. *Fishing for Growth.* Cambridge: Harvard University Press.

Rosenberg, N. 1963. "Technological Change in the U.S. Machine Tool Industry, 1840–1910." *Journal of Economic History* 23 (December): 414–43. Reprinted in *Perspectives on Technology,* chap. 1. London and New York: Cambridge University Press, 1976.

Solow, R. 1956. "A Contribution to the Theory of Economic Growth." *Quarterly Journal of Economics* 70 (February): 65–94.

Stewart, F. 1978. *Technology and Underdevelopment.* New York: Macmillan.

Teitel, S. 1981. "Towards an Understanding of Technical Change in Semi-industrialized Countries." *Research Policy* 10 (April): 127–47.

Teubal, M. 1971. "Development Strategy for a Medium-sized Economy." *Econometrica* 39 (September): 773–96.

Teubal, M. 1973. "Heavy and Light Industry in Economic Development." *American Economic Review* 63 (September): 588–96.

Teubal, M. 1980. "Exportacions de bienes primarios y desarrollo economico." Cuaderno no. 103, Serie ocre, Economia, Institute Torcuato DeTella. ["Primary Exports and Economic Development: The Role of the Engineering Sector." Maurice Falk Institute for Economic Research in Israel, Discussion Paper no. 7915 (1980).]

Uzawa, H. 1964. "On a Two-Sector Model of Economic Growth, II." *Review of Economic Studies* 30 (June): 105–18.

Watkins, M. 1963. "A Staple Theory of Economic Growth." *Canadian Journal of Economics and Political Science* 29 (May): 141–58.

Youngson, A. J. 1959. *Possibilities of Economic Progress.* London and New York: Cambridge University Press.

The author greatly appreciates the comments of N. Gross, G. Hanoch, M. Syrquin, and the participants of the Bar-Ilan Symposium.

# Index

295

COMPOSED BY TSENG INFORMATION SYSTEMS, INC.
DURHAM, NORTH CAROLINA
MANUFACTURED BY CUSHING MALLOY, INC.,
ANN ARBOR, MICHIGAN
TEXT IS SET IN TIMES ROMAN, DISPLAY LINES IN GILL SANS

Library of Congress Cataloging-in-Publication Data
Teubal, Morris.
Innovation performance, learning, and
government policy.
(The Economics of technological change)
Includes index.
1. Technological innovations—Economic aspects.
2. Technology and state.   I. Title.   II. Series.
HC79.T4T485   1987       338′.06      86-40448
ISBN 0-299-10950-X